Rocketdyne: Powering Humans into Space

Rocketdyne: Powering Humans into Space

Robert S. Kraemer

with major contributions by
Vince Wheelock

American Institute of Aeronautics and Astronautics
1801 Alexander Bell Drive
Reston, Virginia 20191-4344
Publishers since 1930

American Institute of Aeronautics and Astronautics, Inc., Reston, Virginia

Library of Congress Cataloging-in-Publication Data on file   1 2 3 4 5

ISBN 1-56347-754-8

Copyright © 2006 by the American Institute of Aeronautics and Astronautics, Inc. All rights reserved. Printed in the United States of America. No part of this publication may be reproduced, distributed, or transmitted, in any form or by any means, or stored in a database or retrieval system, without the prior written permission of the publisher.

Data and information appearing in this book are for informational purposes only. AIAA is not responsible for any injury or damage resulting from use or reliance, nor does AIAA warrant that use or reliance will be free from privately owned rights.

CONTENTS

Acknowledgments, **vii**

List of Abbreviations, **ix**

Introduction, **xi**

1. Rockets, from Theory to the V-2, **1**
2. NAA Chooses the V-2 Path, **15**
3. Navaho I and Redstone with the 75K Engine, **31**
4. Navaho II and III and the 120K/135K Engine, **57**
5. The Atlas with 150K and 60K Engines Orbits John Glenn, **69**
6. H-1 Powered Saturn IB Orbits Apollo 7, Skylab Crews, and Apollo Soyuz, **115**
7. Nuclear Rocket Paves Path to Hydrogen/Oxygen J-2 Engine, **139**
8. F-1s and J-2s Power Astronauts to the Moon, **161**
9. The Fierce Competition for the Space Shuttle Main Engine, **185**

Epilogue, **227**

Appendices, **235**

Notes and References, **253**

Index, **259**

ACKNOWLEDGMENTS

In November 2000 the historians at the Smithsonian National Air & Space Museum (NASM) asked me to come and record my memories of Rocketdyne for their oral history archives. By that time some of the key members of the Rocketdyne team had either died or were in very poor health. With this super-achieving generation passing away, I was concerned that no one had written an overall history of Rocketdyne. The professional historians said there was not enough preserved documentation for them to write a proper history book, and so I determined to visit key survivors of the "glory days" of Rocketdyne to record their memories and save at least some of the Rocketdyne history for the archives. My taping included interviews with Bill Brennan, George Sutton, Bill Cecka, Doug Hege, Bill Ezell, Tom Myers, John Tormey, Paul Castenholz, Sam Iacobellis, Ed Monteath, Steve Domokos, Stan Gunn, Willie Wilhelm, Bob Biggs, Ted Benham, and Vince Wheelock, all past managers of key functions at Rocketdyne. Without exception their eyes would light up when they recalled their experience with the design and development of large powerful rocket engines. Their enthusiasm was so infectious that I determined to make an effort to write a book that would preserve their vital roles in opening the way to the Space Age.

The historians at the space history division of the Smithsonian National Air & Space Museum, especially Roger Launius and Mike Neufeld, gave me encouragement and support in this effort. Mike Neufeld's advice, encouragement, expert knowledge of space history, and detailed critique of my manuscript were invaluable.

For the early history of rocketry up through the work of Robert Goddard in the early 1940s, I have drawn particularly on the history books of T. A. Heppenheimer and Frank Winter, which I have found to be consistently reliable. Other sections of my text where I do not cite specific references are drawn from my own experience and memory. For this book I have the advantage of having been a member of the Rocketdyne team and can describe the evolution of engine design and the contributions and characteristics of key Rocketdyners from my own personal experience with them. Except where I quote others, my assessments are based primarily on my own observations and judgment.

While Rocketdyne had seen no reason to preserve old records, not encouraging to professional historians, there is a marvelous private library of

Rocketdyne history in the home of Rocketdyne retiree Vince Wheelock and his wife Gail. During his 38 years of conducting and managing field services at Rocketdyne, Vince started collecting all forms of documentation, books, newspaper articles, photographs, organization charts, and anything else that recorded a significant event at Rocketdyne. Most of his collection originated years ago in either Rocketdyne or NASA, but would be difficult or impossible to locate today. Gail is a marvelous organizer and has everything nicely labeled and filed, including electronically. Vince now works part time at Rocketdyne organizing historical displays and serving as the closest thing to a curator that they have. His energy is amazing, with enthusiasm to match. He has collected and labeled all of the illustrations in this book and has contributed invaluable details in the text. This book would not have been possible without him—and he is just a great guy to work with.

Vince got valuable assistance in gathering material from current Rocketdyne team members Deanie Snell, Peggy Correa, John Halchak, Maryellen Vetter, Betty Mclaughlin, and Bob Biggs, plus retirees Paul Kisicki, Don Jenkins, Ernie Barrett, Harry Dodson, Phil Fons, Willie Wilhelm, and Norm Reuel (now deceased).

The hard work of getting this book from draft into print was done by the able publishing crew at the AIAA led by Rodger Williams and Alex McCray. Susan Hinmon did an especially nice job in creating an attractive cover.

Finally, last but far from least, I want to thank my spirited Irish-descended wife, Anne, for encouraging me on this book and keeping me well-fed, happy, and healthy. Love you, Anne.

Robert S. Kraemer
August 2005

LIST OF ABBREVIATIONS

AAF Army Air Force
ABMA Army Ballistic Missile Agency
AEC Atomic Energy Commission
AEDC Arnold Engineering Development Center
ARDC Air Research and Development Command
ARPA Advanced Research Projects Agency
ARS American Rocket Society
BMD Ballistic Missile Division
BuAer Bureau of Aeronautics
Caltech California Institute of Technology
DOD Department of Defense
DFRF Dryden Flight Research Facility
EAFB Edwards Air Force Base
E-D expansion-deflection
ERB Edwards Rocket Base
EVA extra vehicular activity
FLOX fluorine/liquid oxygen mixture
FPL full power level
GALCIT Guggenheim Aeronautical Laboratory, California Institute of Technology
GAO Government Accounting Office
HTS horizontal test stand
ICBM intercontinental ballistic missile
IRBM intermediate range ballistic missile
IR&D Independent Research and Development
ISS International Space Station
ISTB integrated subsystem test bed
JATO jet assisted takeoff
JPL Jet Propulsion Laboratory
JSC Johnson Space Center
KSC Kennedy Space Center
LACE liquid air cycle engine
LASL Los Alamos Scientific Laboratory
LLL Lawrence Livermore Laboratories
LM Lunar Module
LOX liquid oxygen
MACE Missile and Control Equipment Division
MAF Michoud Assembly Facility
MEC main engine controller
MIT Massachusetts Institute of Technology
MOC method of characteristics
MSFC Marshall Space Flight Center
MTF Mississippi Test Facility
NAA North American Aviation, Inc.
NACA National Advisory Committee for Aeronautics
NAR North American Rockwell
NASA National Aeronautics and Space Administration
NASM National Air & Space Museum
NERVA nuclear engine for rocket vehicle application
NFL Nevada Field Laboratory
NSTL National Space Technology Laboratories
NTO nitrogen tetroxide
OART Office of Advanced Research and Technology
OMB Office of Management and Budget
ONR Office of Naval Research
OSHA Occupational Safety and Health Administration
RCC rough combustion cutoff device
REAP Rocket Engine Advancement Program
RFNA red fuming nitric acid

RMI Reaction Motors, Inc.
RPL rated power level
SNPO Space Nuclear Propulsion Office
SSC Stennis Space Center
SSFL Santa Susana Field Laboratory
SSME space shuttle main engine
STL Space Technology Laboratories
TRE Test and reliability equipment
TRW Thompson Ramo Wooldridge
UDMH unsymmetrical dimethyl hydrazine
USAF United States Air Force
USSR Union of Soviet Socialist Republics
VfR Verein fur Raumschiffahrt
VIP very important person
VTS vertical test stand
WAFB Wright-Patterson Air Force Base
WDD Western Development Division
WSPG White Sands Proving Ground

INTRODUCTION

At the end of World War II, liquid propellant rocketry in America was still in its infancy, barely out of the cradle. This was surprising in view of the pioneering work of an American, Robert Goddard, who had launched the world's first liquid propellant rocket back in 1926. Nevertheless, by 1945 there were only two small companies in the United States dedicated to the development of liquid propellant rocket engines—Aerojet Engineering, founded by Theodore von Karman in California, and Reaction Motors, Inc. (RMI) in New Jersey, founded by a group of members of the amateur American Rocket Society. They were typically working on pressure-fed rockets with thrust levels of no more than 1500 lb, intended for application on aircraft as jet assisted takeoff (JATO) units. Four of the RMI 1500-lb motors would later be clustered to power the X-1 airplane through the sound barrier. There were also some small research and technology groups here and there, like at M. W. Kellogg, Bell Aircraft, and General Electric, and at government laboratories like the Jet Propulsion Laboratory (JPL) and NACA Cleveland—but that was about it. Government and corporate funding was modest at best and progress was leisurely.

Then seemingly out of nowhere, bursting on the scene like a shooting star, appeared a group at North American Aviation, Inc. (NAA) in Inglewood, California. In 1945 they started small indeed, acquiring JATO units from Aerojet and spare hardware from JPL and firing the motors with a wide variety of propellant combinations in their parking lot at Los Angeles Municipal Airport. This experimental effort would not have been especially notable except that they had much bigger goals in mind and had full corporate backing and encouragement to take bold steps. Aiming at rockets for long-range guided missiles, they made the propitious decision to study and duplicate the breakthrough 56,000-lb thrust rocket propulsion system of the German V-2 missile. This jump to a higher thrust level let them leapfrog over Aerojet and RMI. The NAA version of the V-2's engine was rated at 75,000 lb and was reliable enough to send astronauts Alan Shepard and Gus Grissom into suborbital space. Soon to be named "Rocketdyne," this new team was off and running.

Next in rapid succession came engines of 120,000–150,000 lb thrust for the Navaho, Thor, Jupiter, and Atlas missiles, with the latter propelling astronaut John Glenn to become the first American to reach Earth orbit. Then came the 165,000–205,000 lb thrust H-1 engines for the Saturn I and Saturn IB launch vehicles that put into orbit the first Apollo astronauts, as well as crew members for the large Skylab space station, and then carried American astronauts to an historic handshake in orbit with Russian cosmonauts. By the 1960s Rocketdyne engineers had advanced the thrust level to an almost unimaginable 1,500,000 lb with the giant F-1 engine, still the largest individual liquid engine ever developed. American rocket engines had been advanced in thrust by a phenomenal factor of 1000 times in just a decade and a half.

The combination of the powerful and 100% reliable F-1 engines coupled with high-energy hydrogen/oxygen upper-stage J-2 engines launched all of the Apollo astronauts on their way to the moon. The space race with the Soviet Union had been won in dramatic fashion. Then Rocketdyne developed the far more sophisticated and fully reusable space shuttle main engines (SSMEs) that have been busy ever since 1981 launching hundreds of Americans and their guests into orbit. All of this was accomplished by that one cohesive and focused team.

How does one account for this meteoric rise of a late-comer to so dominate the American liquid rocket engine scene and to advance the dream of sending humans to Mars out of the realm of fantasy and into the beginning of serious planning? In this book we shall try to highlight the reasons for that success story in rapidly advancing rocket propulsion. Our coverage will be focused. Rocketdyne has developed so many small rockets, both liquid and solid propellant, that it would take several thick volumes to adequately describe them all. For example, if you include reaction control, retro, and propellant settling systems, each and every Apollo flight to the moon employed dozens of rocket reaction thrusters from Rocketdyne. We will not attempt to cover all of those. Instead we will focus primarily on Rocketdyne's large liquid propellant engines (thrust levels of 60,000 lb and greater). It turns out that all of those, in one variation or another, have been employed to propel humans into space. In fact, except for the 10 manned Gemini missions, every single human riding American launch vehicles has been powered into space by Rocketdyne engines. That record has stood for over 40 years. Rocketdyne has been a major factor in advancing the United States from trailing the Soviet Union in the 1950s and 1960s into today's position of leadership in space—truly a legendary performance by an exceptional group of rocket pioneers.

While we will be following the saga of technology development, every effort will be made to describe the history and technology in common, generally understandable terms. Much of the story is about the people who made this history, and so the reader will not need any particular technical aptitude to appreciate the interplay of sometimes complex personalities. There will be many heroes but, yes, even a villain or two. We will note a few careers needlessly cut short, but mostly this is a chronicle of triumphant success in the human venture into space.

Rockets, from Theory to the V-2

This book is about humans venturing into space and how they got there. It is a story that begins at least as far back as the twelfth century when the Chinese developed the first rockets, powered by black powder. They were not terribly efficient but made for great fireworks and also short-range artillery rockets. According to rocket historian Willy Ley, Chinese writings record that about 1500 A.D. a wealthy Chinese gentleman named Wan Hoo built a lightweight seat to be suspended from two large kites, beneath which he attached 47 large black powder rockets. He had 47 coolies simultaneously light the 47 rockets, which erupted in a large flash of flame and smoke. History does not record whether there was enough found of Wan Hoo to bury.[1] While he did not succeed in making the first flight into space, one could say that he did make a fast trip into the next world.

It was not clearly recognized for centuries that rockets did not need to push on air and could produce as much thrust (more, in fact) in space than they did at sea level. Writer Jules Verne in his popular classic *De la terre a la lune* (*From the Earth to the Moon*) published in 1865 had his astronauts seated in a large hollow artillery shell that was fired out of a giant cannon. Technically that was impossible—the acceleration alone would have flattened them like a pancake even before atmospheric heating from the extreme muzzle velocity baked them to a crisp. On the other hand, a rocket is a reaction device that accelerates the gas products of combustion and thereby generates a reaction force in the opposite direction. This is in keeping with Sir Isaac Newton's third law of motion that states that for every action there is an equal and opposite reaction.[2] Because the exhaust gases can be expanded and thereby accelerated to even higher velocities in a vacuum, a rocket engine is actually more efficient in space than it is at sea level. Thus a rocket engine could propel a vehicle smoothly through Earth's atmosphere and then continue to accelerate on out into space.

Rocket Pioneers

The first person credited with putting this in solid technical terms was Konstantin Eduardovich Tsiolkovsky, a mathematics teacher born in Russia

2 Rocketdyne: Powering Humans into Space

Wan Hoo Liftoff.

in1857, who taught in the remote Russian village of Kaluga. He developed the basic mathematics of rocket-propelled spaceflight in the late 1800s and published it in 1903 as a paper in the Russian *Scientific Review* under the title *A Rocket into Cosmic Space*. In this paper and in his later work, he described in some detail a rocket for carrying humans into space. In the design he recognized the advantage of using multiple stages for the launch vehicle and achieving high performance by utilizing liquid oxygen and liquid hydrogen as propellants. His work received little attention at the time in Russia and none in western countries.

    The next big step in the mathematics of rocket-propelled flight came from Robert Goddard, a professor at Clark University in Worcester, Massachusetts. Born in Worcester on 5 October 1882, Goddard had a "vision" of traveling to Mars one day while, as a teenager, he was climbing a cherry tree in his backyard. The date was 19 October 1899, and it made such an impression on him that he celebrated its anniversary every year.[3] From that day on he began to analyze means of spaceflight. What was unique about his work is that he did actual experiments, not just analysis, starting with solid propellant rockets. His work earned him a research grant in 1917 from the Smithsonian Institution, which resulted in a 1919 Smithsonian report titled "A Method of Reaching Extreme Altitudes".[4] It outlined the design of a two-stage solid propellant sounding rocket, and then as an example of the potential of such rockets stated that it would be possible to send to the moon a small explosive charge whose flash on impact could be seen from Earth. Of course, that is what drew the attention of the news media, whose coverage was not always complimentary. The *New York Times*, in its 13 January 1920 issue, printed an editorial in which it stated that anyone with even a high school knowledge of physics knew that a rocket needed air to push against to develop any thrust. The *Times* needed to study Newton's Third Law of Motion. Goddard, a shy and reserved man, was very disturbed by this ridicule and determined to minimize any public disclosure of his work. If he wanted to disclose more, he could have reported his test of a small rocket motor in a vacuum tank where it produced 20 % more thrust that at sea-level pressure. By 1920 he had focused his efforts toward liquid propellants, which promised greater performance and thrust control than solid propellants.

In company with Tsiolkovsky and Goddard, the third key pioneer in rocket-powered spaceflight was Hermann Oberth, born on 25 June 1894 in Transylvania, which later became Romania. A brilliant student, he studied physics at the Universities of Cluj (Romania), Munich, Goettingen, and Heidelberg. Inspired from age 11 by Jules Verne's *From the Earth to the Moon*, he applied his knowledge of physics to spaceflight, producing in 1923 a book titled *Die Rakete zu den Planetenraumen* (*The Rocket into Planetary Space*).[5] In only 87 pages it compressed an amazing depth of detail on almost all major aspects of spaceflight, including design features of a two-stage rocket employing liquid oxygen and an alcohol/water mix for the first-stage rocket engines and liquid oxygen/hydrogen for the upper stage. Oberth's work is all the more remarkable for the fact that he was unaware of the work of Tsiolkovsky and Goddard until after his book was published. In spite of its multitude of complex equations, Oberth's book proved extremely popular and sparked the formation of rocket societies in several countries. Oberth expanded his work in even more detail with an edition of more than 400 pages in 1929 titled *Wge zur Raumschiffahrt* (*Ways to Spaceflight*). Smithsonian space historian and curator Frank H. Winter summarizes the relative contribution of Hermann Oberth as follows: "Tsiolkovsky and Goddard technically started earlier and could claim many priorities, but their earliest work remained unknown to the public. Thus, by virtue of his thoroughness and the fact that his ideas were openly published and sparked a wide-spread movement, Oberth alone deserves the title 'Father of the Space Age'."[6] (More details on the works of Tsiolkovsky, Goddard, and Oberth and the early development of liquid rockets are provided in the Appendix of this book. For further detail an excellent source is Frank Winter's book *Rockets into Space*.)

By 1923 the theoretical feasibility of liquid-propellant rocket-powered spaceflight was solidly established and interest was building. It was time to convert theory into hardware and actual flights. Amateur rocket societies began firing small rocket motors but, although cloaked in privacy, it was Robert Goddard who would take the lead. In a field on his Aunt Effie's farm in Auburn, Massachusetts, on a memorable 16 March 1926, he prepared to launch his first rocket. It was a spindly thing standing three meters tall that burned liquid oxygen and gasoline to develop a thrust of only 9 lb (40 N).[7] With his young wife Esther standing by with a stopwatch and clipboard, Goddard directed his technician to ignite the rocket by swinging in a blow torch on a long pole. Rising to an altitude of only 12.5 m (41 ft), it crabbed sideways for a total distance of 56 m (184 ft)—not terribly impressive but still the world's first ever flight of a liquid-propellant rocket.

No one had flown a liquid-propellant rocket before, and so there was no established configuration. Goddard placed the rocket motor (more commonly called the thrust chamber) in the nose of the vehicle, firing down on an asbestos-protected pointed propellant tank. He likely was thinking that such a "pull" arrangement would stabilize the vehicle like an arrow in flight. However, an arrow's feathers only stabilize it because they are immersed in the surrounding airstream. Goddard's tanks in the rear of his rocket were immersed in the motor exhaust and could not feel the airstream. Not only was there no stabilization, but any slight eccentricity in the flow over the tanks would create a side force to cause the rocket to veer, which it did on that first flight. All subsequent Goddard rockets located the thrust chamber in the tail.

The world would have to wait for the announcement of that first flight. And wait and wait. It was not until 10 years later in March 1936 that Goddard submitted to his Smithsonian sponsors his second progress report titled *Liquid-Propellant Rocket Development*.[8]

Pre-World War II Developments

Moving his operation to remote Roswell, New Mexico, in 1930, Goddard made impressive progress. With the help of a supportive Charles Lindbergh, he received sufficient funding from the Guggenheim Foundation to continue his development work. With this support Goddard made important strides in features that would become standards for rockets, like streamlining, aerodynamic fins, automatic remote control, gyro stabilization, steering by air vanes and jet vanes, gimbaled steering, regenerative cooling, and centrifugal propellant pumps driven by a gas generator. He conducted 31 flights in New Mexico, reaching 2.3 km (7500 ft) altitude and speeds in excess of 1127 km/h (700 m/h). One of his latest rockets from 1941 stands in the National Air & Space Museum in Washington, D.C. An impressive 6.7 m (22 ft) tall, it weighs 448 kg (985 lb) fully fueled and incorporates many of the advanced features listed previously.

Unaware of Goddard's progress, the amateur rocket societies of the world tried their own ideas on a small scale. One of the first such groups was the Society for the Study of Interplanetary Travel founded in the Soviet Union in 1924. Other groups formed there, some with government ties, and gained valuable experience for future Russian spaceflight leaders like Valentin P. Glushko and especially Sergei P. Korolev, who survived a Stalin purge that sent him to Siberia to return to lead the monumental Soviet space effort. The amateur society that would have future impact on the United States was the German Verein fur Raumschiffahrt, or VfR (Society for Spaceship Travel), founded in July 1927. Their technical progress was well behind that of Robert Goddard, but one of their most enthusiastic and technically productive members was a young teenager named Wernher von Braun, who in the years ahead was to have a major impact on spaceflight and putting Americans into space.

The principal rocket society formed in the United States was the American Interplanetary Society (in 1934 renamed the American Rocket Society, or ARS). On a minimal budget but making clever use of scrap material, they

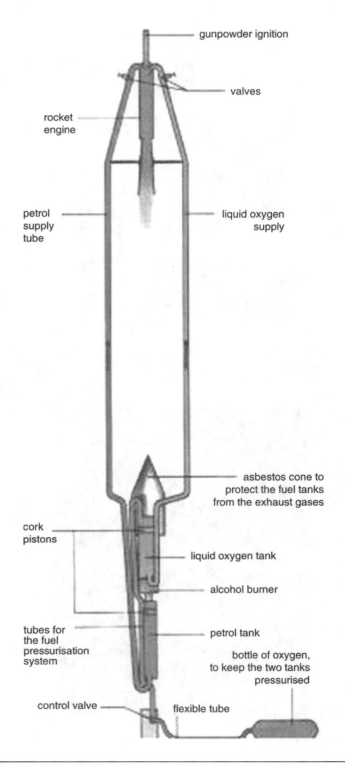

Goddard's first liquid-propellant rocket, 1926.

Goddard launched the world's first liquid-propellant rocket on 16 March 1926, in Auburn, Massachusetts.

Goddard at work on a liquid-fuel rocket in his laboratory at Roswell, New Mexico, November 1935.

began rocket static firing tests in November 1932. Their first flight was from Staten Island, New York, on 14 May 1933, and it reached an altitude of 76.2 m (250 ft). One of their successful developments was a quite reliable regeneratively cooled thrust chamber. In December 1941, immediately after the attack on Pearl Harbor, four of the ARS members, James H. Wyld, John Shesta, Lovell Lawrence Jr., and H. Franklin Pierce, founded Reaction Motors, Inc. (RMI), America's first commercial company for the development of liquid-propellant rocket engines. They are perhaps best known for their later development of the four-barrel oxygen/ammonia 6000-lb thrust engine that in 1947 propelled the Bell X-1 aircraft past the speed of sound.

In addition to Goddard's development activities and the ARS testing, there was a group of graduate students at the California Institute of Technology (Caltech) studying under the legendary aerodynamicist Theodore von Karman and interested in rocket propulsion, especially as a way to propel sounding rockets for high-altitude research. Led by Frank Malina, the group included William Bollay, John W. Parsons, and Edward S. Forman. Von Karman helped them set up a rocket research project under the Guggenheim Aeronautical Laboratory, California Institute of Technology (GALCIT) Program at Caltech, and the group started testing small solid-propellant motors in an isolated dry gully (the upper terminus of the canyon known as the Arroyo Seco) north of what is now the Pasadena Rose Bowl. John Parsons became the propellant specialist and came up with an asphalt binder mix with potassium perchlorate that produced solids with much improved physical properties, like no cracking in cold weather or sagging in hot weather.

The Army Air Corps became interested in their work and funded them to develop solid-propellant jet assisted takeoff (JATO) rocket units to help military aircraft takeoff with heavier loads from short airstrips and from aircraft carriers. This work was so successful that von Karman in 1942 founded the Aerojet Engineering Corporation, located first in Pasadena[9] and then moved to their more remote test site in nearby Azusa, to further develop and then produce the JATO units. Aerojet would soon broaden into liquid-propellant rockets. A year later in 1943 von Karman organized the Jet Propulsion Laboratory (JPL) under the administration of Caltech and located at the head of the Arroyo Seco, where it would build a facility to continue research in rocket propulsion, especially with liquid propellants. The Army pledged support, and Frank Malina was named the first Director of JPL. From research JPL progressed into design and development of rocket-powered vehicles for the Army, notably the Corporal missile propelled by a pressure-fed motor burning red fuming nitric acid (RFNA) and aniline to produce an impressive takeoff thrust of 20,000 lb.[10]

The Caltech/Aerojet success with solid-propellant JATO units sparked interest in liquid-propellant JATOs that could be refueled and reused. The Navy's Bureau of Aeronautics (BuAer) set up a development program in its Power Plant Development Branch in Annapolis under the direction of an enthusiastic young crewcut lieutenant, Bob Truax. Robert Goddard volunteered his expertise and conducted his own parallel development program in Annapolis until his death to cancer in 1945. Organizationally Goddard worked under Truax, but Truax later complained that Goddard declined to trade test results or to share his experience in rocket design.[11] In any event, neither of their liquid JATO units were fully developed or utilized by the armed services.

The Technical Leap to the German V-2

By the end of World War II in the United States, there were only two small commercial corporations devoted to serious development of liquid-propellant rocket engines, Aerojet and Reaction Motors, plus an Army-funded laboratory, JPL, and some Navy-funded work in Annapolis. Other than a smattering of scattered experimental work by small groups, such as those being formed at Bell Aircraft, M. W. Kellogg, NACA Cleveland, and General Electric, that was it for the United States. There was no high priority or major funding for any of these groups. Such low-level effort in America had not been the situation in Germany, which sprang a technical marvel on the world with the first launches of their V-2 ballistic missile on Antwerp and London in late 1944. The V-2 was at that time an enormous leap forward in both size and performance. It was the first vehicle to penetrate out of Earth's atmosphere, and one could well argue that the space age started right then. It was the V-2 that was soon to lead to the formation of Rocketdyne, and so it is important that we briefly review its technical evolution.

Rocket development in Germany had been started by amateurs, whose lack of engineering skills not infrequently led to fatalities. On 19 April 1930 Max Valier successfully tested a rocket car propelled by a motor burning liquid oxygen and water-diluted alcohol. He was killed a month later when that same car exploded, but he had indeed demonstrated that liquid-propellant rockets did work, at least sometimes. That drew the attention of Col. Karl Emil Becker, Chief of the Army's Department of Ballistics and Munitions, and Capt. Walter R. Dornberger, who was developing solid-propellant artillery rockets for Becker. During World War II the Germans had developed the so-called Paris Gun with an extraordinarily long barrel that had a range up to 80 miles, but the gun was too heavy to move in one piece and could only launch a small projectile. The Treaty of Versailles had neglected to mention rockets in its list of weapons forbidden to Germany, and so the German Army focused on developing artillery rockets. Liquid propellants promised much greater range than did the solid propellants of the 1930s, and so Becker summoned Dornberger and gave him written orders that stated, "You have to develop a liquid rocket which can carry more payload than any shell presently in our artillery [and] . . . farther than the maximum range of a gun."[12]

Dornberger gave a contract to Paul Heylandt's semiprofessional group, the Association for the Utilization of Industrial Gases, who had done some rocket motor development work for Valier. However, he had also monitored the work of the amateur VfR and was especially impressed by the outstanding technical abilities and poise of one of the young VfR members, Wernher von Braun. In October 1932 he hired von Braun as his technical assistant and began setting up a rocket research facility at an old Army

Rockets, from Theory to the V-2   **9**

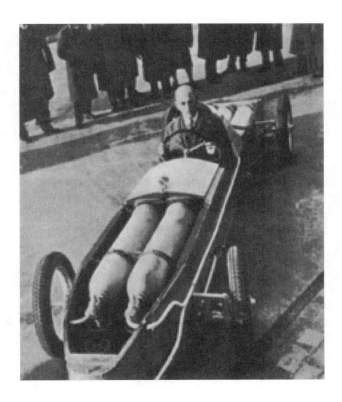

Max Valier in his liquid-propellant rocket car.

range near Kummersdorf, south of Berlin. Their first rocket vehicle, the A-1, which stood 1.4 m (4.5 ft) tall, was not a great success, but its successor A-2 with improved gyro stabilization reached an altitude of about 2 km (6500 ft).

Buoyed by this success, Dornberger and Becker were able to negotiate for major funds from both the Army and the Luftwaffe (who wanted a rocket engine for their ME-163 interceptor aircraft) to build a large rocket development facility near the tiny fishing village of Peenemunde on the island of Usedom on the Baltic Sea. The remote site promised secrecy, and the Baltic gave them a long-range testing area. Dornberger demonstrated his great confidence in the abilities of von Braun by naming him, at just age 23, as the technical director of the Penemunde operation.

Undaunted by stabilization failures of all four of their 6.5-m (21.3-ft) tall A-3 rockets, they charged ahead for the big one, the A-4. Dornberger wanted a payload of one ton of high explosives to be carried to twice the 80-mile (129-km) range of the Paris Gun, the "ultimate" artillery piece. One constraint was that the A-4 had to be carried on freight cars through standard railroad tunnels. The final design had a diameter of 1.65 m (5.4 ft) and a height of 18.5 m (46.9 ft). That may not be impressive today, but in 1937 it was enormous and ambitious almost beyond reality.

The chief designer was von Braun's deputy, Walter J. H. Riedel, known as "Papa" by the Peenemunde crew. He drew a streamlined shape, stabilized by large fins and guided by gyros driving graphite jet vanes immersed in the rocket engine exhaust. The structure was quite conservative, with individual propellant tanks mounted inside the missile skin. The real challenge was the rocket engine, much larger than any before attempted. Various sources give different numbers for the design thrust level (nominally 25 metric tons), with experts like Frank Winter[13] and Willy Ley[14] putting it at 59,500 lb, while engineers at North American Aviation (NAA) always considered it to be 56,000 lb. In any event, it was more than an order of magnitude greater than had ever been attempted before. Responsibility for the new engine design was given to chemist Walter Thiel.

The selected propellants were liquid oxygen and ethyl alcohol diluted with 25% water to ease thrust chamber cooling. To meet the required weight, the propellants would have to be pump-fed rather than coming from pressurized tanks. The required pump technology came from a surprising source, the light-weight centrifugal pumps used by firefighters. The resulting A-4 design used an inline coupled-two-piece drive shaft with a two-stage turbine wheel in the center, the oxygen centrifugal pump on one end of the shaft, and the alcohol pump on the other end. The turbine was driven by steam generated from decomposing hydrogen peroxide. That decomposition in the gas generator was achieved with a potassium permanganate catalyst, but even trace contaminants could also trigger decomposition, so that one had to be very careful to maintain the purity of the hydrogen peroxide.

Small-scale tests indicated that a good volume in the combustion chamber was required for complete burning, and so the A-4 chamber was almost spherical and encompassed a generous volume. Thiel ran into difficulty making a large propellant injector, and so he compromised by employing a cluster of 18 of the small cup-shaped injectors that had given good combustion efficiency in the A-3 motor. It worked, although it looked awkward as there were 18 oxidizer lines leading to the injectors—like a tangle of giant squid. The next problem was the hot spots that developed in the chamber walls just downstream of the injectors. The simple solution was to drill holes in the walls at the hot spot locations so that alcohol could stream out to provide film cooling. The engine was started slowly with a low-level prestage, and combustion stability was never a problem.

There were failures in the static firing tests and in the first two flight attempts. There is a much screened documentary film from Peenemunde of an A-4 falling back to the ground and generating an enormous fireball. However, the third flight test on 3 October 1942 was a complete success, reaching a speed of 5300 km/h (3,300 miles/h), an altitude of 97 km (60 miles), and a range of 200 km (125 miles). At the triumphal celebration that evening Dornberger hailed the opening of a new era of "space travel," but then he reminded the Peenemunde workers that they still had a weapon to deliver. Based on that success, Hitler gave the A-4 the new name of Vengeance Weapon 2, or V-2 (the predecessor V-1 was a pulsejet-powered subsonic cruise missile), and gave it highest priority. The Nazi SS set up an underground production facility, Mittelwerk, using mostly slave labor, and produced more than 6400 V-2 missiles, which began raining down on London and Antwerp. Fortunately it was late in 1944 and Germany's war was about to end, no matter the damage from the V-2s.

As U.S. troops swarmed into Germany, Army Col. Holger N. Toftoy was really on the ball. He located Wernher von Braun and his key people, who were waiting in Oberammergau in southern Germany for the U.S. troops to arrive—they were determined not to be captured by the hated Russians and thought that America rather than Great Britain was where future space travel would be pursued. Colonel Toftoy raced trucks to Mittelwerk and gathered parts for about 100 V-2s for shipment to the United States, where many would be assembled and launched as sounding rockets from the White Sands Proving Ground (WSPG) in New Mexico. With guidance from the von Braun crew, Toftoy was also able to locate and recover tons of design documents that they had taken with them out of Peenemunde and hidden away before the Russians arrived.

Von Braun and his select team of 118 specialists were escorted as "guests" of the Army to Fort Bliss in Texas, close enough to WSPG to supervise the assembly and launch of the captured V-2s. As that effort wound down in 1950, the Army transferred von Braun and his German crew to the Redstone Arsenal in Huntsville, Alabama, and set them up as the core of the Army Ballistic Missile Agency (ABMA). Some eyebrows were raised at the Army's embracing their former "enemies," and a number of them, including von Braun, were investigated for their membership in the Nazi Party, but only one, Arthur Rudolph, was ever accused of war crimes. Soon after NASA was formed in 1958, von Braun's design and development team at ABMA, most of whose members were now U.S. citizens, became the Marshall Space Flight Center, which has led NASA's development of launch vehicles ever since.

As we shall see, the V-2 was an all-important stimulus to aerospace technology in America. Wernher von Braun turned out to be an extremely valuable asset. His political role in Germany is still debated, e.g., Frederick Ordway III and Mitchell Sharpe, in their book *The Rocket Team*,[15] totally accept the Dornberger and von Braun accounts of how the Nazi SS kept accusing von Braun of stalling and even arrested him for being more interested in space travel rather than developing weapons. On the other hand, the respected aerospace historian Michael Neufeld, who has made an in-depth study of von Braun in his book *The Rocket and the Reich*,[16] shows the one known photo of von Braun in an SS uniform and faults him for not objecting to the heinous treatment by the SS of the slave laborers building V-2s at Mittelwerk. In this book we will be describing von Braun's important roles in this country, and documenting that the people at Rocketdyne had nothing but admiration and respect for him as a leader in venturing into space. Even Neufeld describes

V-2 test stand at Peenemunde.

V-2 missile deployment on trains.

him as "charismatic" and as possessing "prodigious quantities of charm, tact, intellect, and leadership ability."[17] Whatever one's views of his political involvement in Germany, no one can deny that the arrival of von Braun and the captured V-2s in the United States marked a major milestone and beginning of accelerated development in American rocketry.

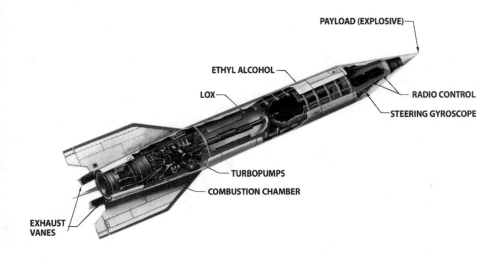

V-2 missile cutaway.

CHAPTER 2

NAA Chooses the V-2 Path

In 1945 at the conclusion of World War II, prompted by the technical advances of the V-2, a new entity appeared in the field of rocket propulsion. Initially a very small group of young engineers, that new entrant would take root and later become renowned as Rocketdyne. Like a newly sown seed, it would only grow if planted in fertile soil. That fertile environment was provided by North American Aviation, Inc. (NAA), a corporation headquartered at the Los Angeles municipal airport in Inglewood, California. NAA was noted for designing and building some of the most successful military aircraft of all time, including trainers, bombers, and fighters. During World War II North American built more than 42,000 aircraft, the most of any American company, with a plane coming off its production lines every 15 minutes, which was truly remarkable. To understand why NAA was so uniquely successful and how it was able to surge into a leadership position in rocket propulsion, it is necessary to go back to its origins and early years.

The founder of North American Aviation, Inc, was an ambitious and imaginative investment banker named Clement Melville Keys, who established the corporation in 1928 as a complex holding company—and "complex" is no exaggeration. Over the next few years it was involved in the creation of Transcontinental Air Transport (TAT), Curtiss Flying Service, Pitcairn Aviation Company, Sperry Gyroscope Company, Eastern Air Lines, Curtiss-Wright Corporation, Ford Instrument Company, Berliner-Joyce Aircraft, Trans World Airlines (TWA), and Western Airlines. How is that for a mix? In 1933 General Motors, which had already acquired Fokker Aircraft (founded by Tony Fokker, designer of the famous German Fokker airplanes of World War I), acquired a 30% controlling interest in North American Aviation and installed Ernest R. Breech as Chairman of the Board.[1,2,3]

The Kindelberger-Atwood Team

In 1934 Breech hired Chief Engineer James H. "Dutch" Kindelberger away from Douglas Aircraft and installed him as President of NAA, which was located in Dundalk, Maryland. Up until then NAA had never built a single aircraft, but Dutch was going to change that in a hurry. Learning that the Army Air Corps was going to hold a competition for a basic trainer aircraft in just three months, he assembled his small but capable crew, many of whom were skilled craftsmen brought over from Holland by Tony Fokker. They not only designed and built an airplane in just nine weeks, but they also flew it to Dayton, Ohio, for the trials at Wright Field. They came home with a contract for an eventual 267 of the trainers, designated the 'basic trainer' BT-19. To produce the planes, they moved the small company out to an airport in Inglewood, California (today's Los Angeles International Airport), where the weather promised more good flying days than did Maryland.

Dutch was the true father of North American Aviation as a company that built airplanes and was one of the most colorful characters in the business. Born on 8 May 1895 in Wheeling, West Virginia, his engineering education was limited to a short period at Carnegie Institute of Technology before he joined the Army when the United States entered World War I. While serving as a second lieutenant in the Signal Corps Aviation Section, he became fascinated by aircraft, and after the war he went to work for Glenn L. Martin's aircraft company in Baltimore, working beside another draftsman named Donald Wills Douglas, who left in 1920 to start his own aircraft company in Santa Monica, California. In 1925 Dutch, who was by then Martin's chief draftsman, accepted an offer from Donald Douglas to join him at Douglas Aircraft Company, where the hard-working young Kindelberger, still in his thirties, was soon vice president and chief engineer. During the design of the pioneering Douglas DC-1, DC-2, and eventual DC-3 transports that really created widespread commercial air travel, Kindelberger noted the exceptional talents of a young structures engineer named John Leland "Lee" Atwood. Thus in 1934 when Dutch was hired by General Motors to be their president of North American Aviation, he hired Lee Atwood away from Douglas as his chief engineer. Their immediate success with the BT-19 trainer was just the first of many triumphs for this outstanding pair.

Of the two partners, Dutch was certainly the more colorful. As he is not here to protest, we can say that he had the face of a cherub, with twinkling eyes, deep dimples, rosy cheeks, and a mouth with upturned corners that almost always seemed ready to break into a broad grin. He was a natural leader who liked people and they liked him. Looks to the contrary, he was described as "crusty" and, as the occasion dictated, could trade language and stories with the roughest dockside stevedores. He made it a point to walk through the shops every day to see actual hardware taking shape, and the assembly workers loved him as one of them. Technically his expertise was in production, where he insisted on design simplicity and ease of manufacture. The resulting producability and low costs made for early success of North American Aviation and proved absolutely vital in meeting the later production demands of World War II.

Behind that cherub face was a firm determination and a quick mind. He knew that the aircraft procurements of the Army Air Force were confusing to

James Howard "Dutch" Kindelberger.

Chairman of the Board Kindelberger with President Lee Atwood, 1961.

contractors because sometimes there was funding for fighters when the need was for trainers and vice versa. After the success of the BT-19 series, Dutch won the next "fighter" procurement by proposing another trainer, which the Army rationalized by calling it the 'basic combat' BC-1 trainer. NAA delivered 300 of those. Next Dutch thought the Army would really need a fighter, or at least that is the story passed around in the company. He gave his sales pitch to the Army, concluding that "this is the best damned fighter ever designed." The ranking general said fine, but what they really needed was an advanced trainer. The story goes that Dutch did not even blink as he responded,

"General, you misunderstood me. I said that this is the best damned trainer ever designed." True story or not, Dutch came home with a contract, and that plane became the famous 'advanced trainer' AT-6 (later shortened to T-6) Texan, the most successful trainer series in history. The original concept of it as a fighter became true later when a modified single-seater armed version was built for overseas sales.

The other member of this close partnership, Lee Atwood, could not be more different from Kindelberger. Lean and serious, appearing to be taller than his six feet, he looked the part of the intellectual that he was. Smiling did not seem to come easily to him. Later in 1963 the publication *Aerospace Management* would say of him, "Lee Atwood gives the impression of being a very modest, retiring, sometimes troubled man, who is constantly aware of the responsibilities on his shoulders. He speaks very softly and slowly, carefully picking his words."[4] Similarly *Forbes* magazine wrote, "Lee Atwood, who both looks and talks like an intellectual . . . a man who goes out of his way to avoid the appearance of boasting . . . runs this complicated show on a pragmatic basis . . . a man not given to flamboyant statements."[5]

Lee Atwood was born in Walton, Kentucky, on 26 October 1904, to a family of English ancestry. His father was a highly respected minister and head of a department at Hardin-Simmons University. The young Lee earned a B.A. degree from Hardin-Simmons followed by a B.S. degree from the University of Texas, majoring in mathematics and structural analysis. His first job in 1928 was with the Army Aircraft Branch at Wright Field. Moving to California, he took a job with Douglas Aircraft Company in Santa Monica, where he made major contributions to the stress analysis and structural design of the milestone DC-1, DC-2, and DC-3 transport aircraft, making a very favorable impression on his boss, Dutch Kindelberger, and leading to his joining Dutch at North American Aviation.

Dutch and Lee were different, but complementary, and together they forged a mighty company. In December 1941 when the United States entered World War II, they had three aircraft already in production—the T-6 Texan advanced trainer, the B-25 Mitchell twin-engine medium-attack bomber, and the P-51 Mustang fighter. The T-6 was already on its way to becoming legendary and was flown by pilots of every Allied air force during World War II. The B-25 became famous overnight on 18 April 1942, when Col. Jimmy Doolittle (already noted as the only pilot ever to tame the world-speed-record-holding Gee Bee racer) led the bombing raid of 16 carrier-launched Mitchells on homeland Japan. The P-51 Mustang in an earlier tamer form was already on order by Great Britain.

When the Allies started intensive bomber raids on the industrial interior of Germany, their planes were beyond the range of escort fighters and attrition was staggering, with losses of over 25% of the planes on some raids. American Air Force generals are reported to have personally asked Dutch Kindelberger if there was anything NAA could do. The best hope was the Mustang. It already incorporated an Atwood-inspired under-fuselage radiator that not only reduced the drag of a conventional frontal radiator but also used the added thermal energy to induce a jet exhaust effect that produced a net thrust rather than drag. The substitution of the powerful 1595 hp Rolls-Royce Merlin engine (produced under license by Packard as the V-1650) in place of the original Allison V-1710 had already made the Mustang the fastest prop fighter of the war, and the fuselage could be modified to add another internal fuel tank. However, the decisive modification was the switch to a laminar flow wing. That feature is so important that it is worth a bit of explanation.

The boundary layer of air passing next to the skin over an airplane's wing starts out smoothly in what is known to aeronautical engineers as laminar flow. Eventually as it passes over the wing, the boundary layer flow piles up and transitions to turbulent flow. This turbulent flow has a beneficial effect in that it mixes the boundary layer with the main airflow, preventing the boundary flow from separating from the surface of the wing, which would be a stall and would kill the wing's lift. The negative part of turbulent boundary layer flow is that it has higher drag than laminar flow. Thus you want turbulent flow during landing, to delay stall, but not during cruise. One of NAA's aerodynamicists, Edward Horkey, had come up with a concept based on research by the National Advisory Committee for Aeronautics (NACA) that would drastically change the cross-sectional profile of the Mustang's wing by moving the point of maximum thickness to well aft of the center of the wing so that the boundary layer flow would continue to accelerate longer and remain in smooth laminar flow. Horkey's design looked like no wing profile ever flown, but in the wind tunnel it promised substantially reduced drag. A laminar flow wing was mounted to a Mustang, and the results were dramatic, with greatly reduced drag during cruise. The takeoffs and landings were a bit more tricky, but cruise was great. Mustangs could now accompany the B-17 and B-24 bombers all the way into Germany and were more than a match for the German Messerschmitt ME-109s. German Luftwaffe chief Hermann Goering was reported to have said that the moment the Mustangs appeared over Berlin the war was over.

During World War II North American Aviation built 42,683 aircraft—a remarkable number and by far the most of any American company. The reputation of the company for delivering superb aircraft on-time and on-budget was firmly impressed on U.S. Air Force (USAF) officers, and that would have a favorable influence on contract awards in the years ahead. However, the immediate impact of the cessation of hostilities in 1945 was negatively dramatic on company employment, which dropped abruptly from 91,000 to 5,000, and there was not really contractual work for that remaining core of 5,000. At that point Kindelberger and Atwood could have paid out the profits made during the war production and pretty well have gone out of the aviation business. They discussed making aluminum products like non-rusting garbage cans, lightweight hospital beds, or even canoes like Grumman was to make. Instead they said, "No, we are an aviation company and we build airplanes." To keep their core aircraft design and fabrication team intact, they elected to build the luxury four-place aircraft, the Navion, even though they knew it would never be profitable.

The key decision here is that they wanted to do challenging engineering work. They considered commercial aviation to be not only financially risky but also not as technically challenging as meeting future military requirements. Their risk assessment of commercial aviation proved correct, as not only did the Navion lose money but so did their later venture with the Saberliner executive jet, a modification of a plane they designed for the USAF. NAA was going to follow in the path of the laminar flow P-51 and continue to pursue breakthrough technology advances, like the record-breaking

North American Aircraft's XB-70 Valkyrie airplane.

North American Aircraft's X-15, the winged rocket record breaker and space prober.

X-15 aircraft. Another good example would be the remarkable XB-70 bomber that would droop its wing tips and utilize shock waves to trap a bubble of high-pressure air under its wing so that it could cruise for long ranges at up to three times the speed of sound. The only other aircraft that would ever match that capability was the extraordinary titanium-skinned Lockheed SR-71 Blackbird.

The Move to Guided Missiles

More importantly for setting their future direction in 1945, Kindelberger and Atwood assessed the results of World War II and judged that in addition to the need for future military piloted aircraft, the German development of the pulse-jet-powered V-1 cruise missile and the rocket-powered V-2 ballistic missile had introduced a new era of the guided missile. To meet future military requirements would require new technology advances in many fields, including supersonic aerodynamics, rocket and jet propulsion, inertial and stellar guidance, high-temperature structures, nuclear reactions, and advanced electronics for automatic control. To build these capabilities, they were prepared to devote their substantial company assets, and they would need the leadership of a skilled technical manager. The man they selected was William "Bill" Bollay.

Born in Germany in 1911, Bollay's family had moved from Stuttgart to Illinois in 1924. He started his college education at Northwestern University but in 1933 earned a scholarship at Caltech, where he would study under the guidance of Theodore von Karman. There he joined the GALCIT crew and teamed with Frank Malina in the testing of small solid-propellant rocket motors. However, his main interest and thesis topic were in aerodynamics. After receiving his Ph.D., he taught for a while at Caltech and then Harvard,

Bill Bollay.

where he built a small wind tunnel for aerodynamic studies. In September 1941 he started active duty in the Navy and was appointed to head BuAer's Power Plant Development Branch in Annapolis. Doing liquid-propellant rocket JATO development in his branch were both Bob Truax and Robert Goddard, but Bollay himself concentrated on the development of turbojets. It was turbojets that led him to working with North American Aviation, which was developing the FJ-1 Fury aircraft as the Navy's first carrier-based jet fighter.

When Kindelberger and Atwood were searching for a technical leader to propel NAA into the guided missile age, Bollay quickly headed the list of candidates. Not only did he have excellent academic credentials but also the NAA engineers knew and respected him. A rather quiet person, he nevertheless radiated leadership. His slim face and heavy dark eyebrows could concentrate with impressive intensity on the problem at hand and inspire that intensity in others. He was hired by NAA in November 1945 and given the second floor of an old tooling building right on Imperial Boulevard at the Inglewood airport, which he soon named the Aerophysics Laboratory. The quarters were nothing fancy—you had to walk through a machine shop on the ground floor and then up a flight of stairs to the loft, which was a big bare room with just the roof's wooden rafters for a ceiling. There was no heat or air conditioning, but fortunately the Inglewood location was favored year round by moderating breezes off the ocean.

Bollay was given the green light to begin hiring the specialists he needed. In this hiring he was very selective, going for quality before quantity. As his brother Eugene was to summarize it, "Bill brought into North American an unusual bunch of highly gifted people. And they were all like him. They all were pushing like mad to be the top in their particular field."[6] This hiring of the cream of the crop enabled the new engineers and scientists to grow in both technical skills and management abilities as the Aerophysics Laboratory began to greatly expand in scope in the many technical disciplines required for developing guided missiles. Propulsion alone required a capability in ramjet and turbojet propulsion as well as rocket propulsion. Bollay assigned Leonard Gore to coordinate this buildup in overall propulsion capability.

Building a Rocket Team

In the field of rocket propulsion there were very few who met Bollay's criteria for both academics and experience. Von Karman's GALCIT graduate program at Caltech was one source, and Bollay and his deputy, Edward Redding, quickly hired George P. Sutton, Thomas F. Dixon, and then Douglas W. Hege from that program. Sutton had not only an academic background in rocket propulsion but also several valuable years of experience as a rocket development engineer at Aerojet Engineering. In 1949 he authored a book called *Rocket Propulsion Elements* that became the primary textbook on rocket engine design and has been greatly expanded in scope and continually in print for over 50 years. Sutton was understandably considered the rocket "guru" of the early group.

Another early hire, John Tormey, was a chemical engineer with experience with hydrogen-peroxide-powered torpedoes, and so he was given the charter to become an expert in liquid propellants. Tormey in turn quickly hired John Parsons, the solid-propellant expert from Frank Malina's Caltech rocket team, and then Stanley Greenfield, a graduate of Michigan State and a specialist in heat transfer, and Stanley Gunn from Zucrow's combustion research group at Purdue. Greenfield and Gunn were each to make important contributions to rocket engine advancements. Then came a group of exceptional mechanical engineers, like Paul Vogt and Norm Reuel (also a Caltech graduate), who had no rocket hardware experience but knew the field of pumps and machinery. A real find was Matt Ek, who was downright brilliant in turbopump design and especially in stress analysis. A bit later they hired Roy Healy, one of the early rocket experimenters and an employee of the Army Air Corps in the development of air-launched solid rockets. Roy had served as president of the American Rocket Society in 1942 and again in 1947, as would George Sutton in 1958. The very intelligent Healy was widely respected and would give valuable guidance to the test and development crews in the years to come.

Tom Dixon was a couple of years older than most of the youngsters and had some large rocket experience, having been assigned to ferret out the V-2 secrets during World War II, and so serving as Len Gore's deputy he was charged with coordinating the rocket propulsion effort. Tom Myers, who had been working on ramjet combustion, was made responsible for rocket research,[7] with John Tormey leading the experimental research testing.[8] A young member of Tormey's test team was Paul Castenholz, hired in 1949, who in years to come was to play a major role in the success of NAA's rocket engine development. While Dixon and Myers were the nominal team leaders, all of these guys were in their mid to late twenties and quickly established an informal and close working relationship, acting as a close-knit team with no one worrying much about job titles or levels of management. Working with rockets was new and exciting, and so job interest and morale was high. The guys played cards together every week, some at poker and others at bridge, depending on their particular tastes and fascination with either the psychology or mathematics of card playing. Among the bridge players Tormey and Reuel were especially good and even won a few large tournaments. Many of these guys were such buddies that they even served as occasional babysitters for each other's small children.

Early Testing

Dixon picked William J. Cecka, whose experience included the development of launchers and firing tests of 4.5-in. (11.4-cm) armament rockets from aircraft, to head up the testing of available liquid-propellant rocket motors in the 300–1500-lb range,[9] while Tormey's research crew was to use tiny motors to

George Sutton   Tom Dixon   Doug Hege   John Tormey

Stan Greenfield   Stan Gunn   Paul Vogt   Norm Reuel

Matt Ek   Roy Healy   Tom Myers   Paul Castenholz

Building an early NAA rocket team. (Note: Unable to locate photographs of Ed Redding and John Parsons.)

explore the performance potential and properties of various propellant combinations. NAA had a large parking lot bounded by Aviation Boulevard and Imperial Boulevard at the northeast end of their Inglewood property, and an area of 163 ft by 250 ft (49.7 m by 76.2 m) was fenced off in this east parking lot for rocket testing. A 10-ft (3-m) tall wall was built bisecting the area, with half devoted to the research testing. Tormey went off to Pasadena and came back with "surplus" thrust chambers courtesy of JPL, settling in on 50-lb motors as an adequate size for propellant screening. These motors were so small that they whistled rather than roared when fired. Cecka, on the other hand, needed to gain experience with larger motors. Aerojet had developed for BuAer two JATO units of 1000 and 1500 lb thrust burning nitric acid with aniline, but these units never went into production. With help from the Army Air Force and BuAer (notably Bob Truax), two of these units were declared surplus and "loaned" to NAA for testing in the east parking lot.

For the first small-motor tests they protected the test crew by using a forklift truck to hoist a scrapper blade from a bulldozer to act as a blast shield between the motor and the crew. The next step up in sophistication was to build a blockhouse. During the war dozens of poured concrete shelters had been located throughout the NAA plant in anticipation of Japanese air attacks on Los Angeles. One of these was hoisted and plunked down in the center of the new test area with viewing slits cut into the walls. The slits were not too reassuring, and so they were replaced by imbedding bulletproof windshields from Mustang fighters. These windshields were double-paned, with each layer 1.5 in. thick, and could stop a pretty good sized piece of shrapnel. They proved to be quite satisfactory, several times stopping flying fragments.

Bill Cecka is a very conscientious man, deeply concerned about safety, and he was beginning to feel more comfortable about east parking lot safety. However, the acid/aniline combination was not easy to work with. Any rare spill of the red fuming nitric acid (RFNA) produced a small cloud of reddish-brown acid vapor that had to be dispersed, and the hardware and test stand after each test had to be immediately flushed with water and a neutralizing solution. Test work began to shift more and more to liquid oxygen rather than nitric acid.

Then came a day in December 1947 that Bill still recalls with a shudder. Everyone on the rocket team found the tests with all of the flame and noise to be exciting, and so the blockhouse was crowded that day with 25 team members for the first test of a newly designed all-steel 300-lb motor. It employed a purchased Aerojet pintle valve assembly that incorporated both fuel and oxidizer valves in one housing. Unknown to the test crew, the valve had leaked fuel and formed a puddle in the combustion chamber. When the valve was actuated the liquid oxygen instantly formed a jell with the puddled fuel. On ignition the motor exploded violently, which was not a rare event in those days. However, what was not realized was that a metal fragment had ricocheted off a side wall and punctured a high-pressure acetylene tank mounted behind the blockhouse. The stream of gas from the punctured tank was strong enough to push open the spring-loaded back door to the blockhouse, allowing the acetylene to blow inside. Now acetylene, when mixed with air, is violently explosive, and if a spark had set it off, there would likely have been 25 dead people. Fortunately one of the engineers, Robert Lawrence, recognized the odor and a characteristic haze from a balsa powder used as a stabilizer in acetylene tanks, and he immediately shouted, "Don't touch any switches!" There was no spark and all escaped, but Bill Cecka still looks haunted when he tells of that event.[10]

After the succeeding test series with the 300-lb motor, a new 1000-lb motor was designed and tested extensively with the same liquid oxygen and 75% ethyl alcohol/25% water propellant combination used by the V-2. By then the test instrumentation had greatly advanced in sophistication, with measured parameters including thrust, flow rates, propellant pressures and temperatures, chamber pressure, chamber and nozzle gas temperatures, wall temperatures, jet gas velocities, and flame characteristics. Test results were used to progressively optimize mixture ratio, chamber length, and injector design. The best performance was achieved with an injector using a triplet pattern, with two fuel streams impinging on each oxidizer stream, resulting in a measured combustion efficiency of 94.8% of theoretical.[11]

Ignorance was bliss in those early pioneering days. Testing was readily approved by the people at the Los Angeles Municipal Airport and the Los Angeles Fire Department, and the local fire station was notified before each firing test. Employee cars were parked just outside the testing area fence, as were Navion airplanes coming off the production lines. Helped by the fact that it was an industrial rather than residential area, there were no complaints about noise or fumes. The test area was paved so that spillage could be readily collected. Water flushes and blowers were provided, and safety was drummed into the test crew.

Yet looking back with today's knowledge of hazards, some of the propellant combinations tested in the east parking lot were pretty wild. Nitric acid and aniline were bad enough, but Tormey's researchers were evaluating the performance of potential high-energy propellant combinations. Hydrazine was tested with liquid oxygen as a step up in performance from the V-2's LOX/ethyl alcohol. A much more bold test series was with liquid fluorine and hydrazine. Just the hydrazine is pretty tricky stuff—it looks like innocent water but as a monopropellant it can be triggered into explosive combustion by impurities, and it is readily absorbed through the skin to cause internal damage—but the fluorine is even worse. It reacts with almost everything, will even burn water, and is highly toxic to breathe in concentrations as dilute as one part per million. Today's Occupational Safety and Health Administration (OSHA) would faint dead away at the thought of testing it at the Los Angeles airport, but in 1946 the Los Angeles Fire Department gave a green light to the test requests. The research crew even considered testing liquid ozone but concluded it was just too unstable to work with. Both Tom Myers and John Tormey consider that the major contribution of the research group in those early days was to *eliminate* some of the more exotic propellants being considered.

What reduced any risk, besides the caution of the test crew and their use of masks and protective clothing, was the very tiny size of the 50-lb motors and the very small quantities of propellants used in each test. Other than a few minor acid burns, there were no reported injuries throughout the testing in the east parking lot. It was probably the day when some shrapnel from a blown motor damaged a nearby car in the parking lot, plus the need to go much higher in thrust levels, that Bollay determined to find a better remote test area.

Building a Remote Test Facility

An excellent site was located in the Santa Susana Mountains at the far west end of the San Fernando Valley. It was an easy drive of just an hour and a half from the Inglewood plant out Sepulveda and Ventura boulevards, and there was no encroaching real estate development up in the Santa Susana hills. The terrain was quite rugged and barren, and large acreage was owned by just one

North American Aviation Los Angeles Aircraft Division, east parking lot early test crew, back row: Bob Holbrook, Unknown, Ed Redding, Stu Warren, Tom Carvey, and Unknown, Front Row: Bill Cecka, Ray Keeler, Unknown, Al Stanley, and Jim Benson.

NAA Los Angeles Aircraft Division, East Parking Lot test facility, 1946–1949.

NAA LAAD parking lot control center, 1947.

family, the Dundas. They occasionally allowed it to be used for the filming of movie westerns—it is an absolute certainty that if you ever saw any early Tom Mix or Lone Ranger westerns, you would instantly recognize the huge boulders and erosion-carved sandstone structures of Santa Susana. The Dundas family agreed to a long-term lease, and in March 1947 Kindelberger and Atwood dug once again into NAA cash reserves for another million dollars, including $713,000 to go along with $1.5 million from the Air Force to develop a test site and the balance to build at Inglewood a blow-down wind tunnel capable of limited-duration tests in supersonic flow. This funding was just one of many examples where Dutch and Lee did not hesitate to allocate company resources to elevating the technical capability of the Aerophysics Laboratory.

While the Santa Susana facility was being prepared, the level of testing in the east parking lot was elevated to testing a new motor of 3000-lb thrust. Testing of this motor design would then be continued at the new site after test operations in the east parking lot were concluded in August 1949. Bill Cecka played a leading role in the design of the new test facility, with strong support from Doug Crossland, Joe McNamara (who was to rise rapidly in the organization), Bob Lodge, and Duncan Jolicoeur.

The Pioneering MX-770 Navaho

Meanwhile back in November 1945, the Army Air Force (AAF) invited contractors to submit proposals on several categories of missiles. Bill Bollay made the decision to bid on the field of surface-to-surface guided missiles and, working with NAA chief engineer Ray Rice, made the all-important decision that they would use the V-2 as their starting point for guided missile design and development. The Germans had attached wings to their A-4, calling it the A-9, making it a boost-glide vehicle as a way to extend its range. While the wings failed in a flight test, the concept seemed sound, and NAA so briefed AAF personnel. On 22 April 1946, the AAF awarded NAA a letter contract for $2.3 million to start design of a supersonic boost-glide guided missile with a range of 500 miles. The project would be designated MX-770.[12] North American management had an inclination toward names that began with their corporate initials NA, and so they appropriately named the new missile Navaho I. Some people say the initials came from the rocket propellants used, i.e., "North American Vehicle, Alcohol, Hydrogen peroxide, Oxygen." It was noted that the Navaho Indians once lived in huts called "hogans," and so the field shelters that were to house the operational Navaho missiles were dubbed "Navahogans."

In December 1947 NAA leased the former Chance Vought facility and airstrip on Lakewood Boulevard in Downey, California, for production of T-6 and T-28 trainers. Soon the growing Aerophysics Laboratory was relocated there and in 1948 was given the new name of Missile and Control Equipment (MACE) Division. The Downey plant was no plush facility, with cavernous large rooms, no ceilings or insulation between floors and bare roofs, and no air conditioning to ease the hot summers. Moreover, across the airstrip to the east was the Red Star fertilizer plant with its huge mounds of cow and horse manure. Fortunately

Bill Cecka   Doug Crossland   Joe McNamara

Bob Lodge   Duncan Jolicoeur

Key personnel in design of remote test facility in Santa Susana mountains.

the prevailing wind was out of the west, but during the occasional periods of Santa Ana winds out of the east, the odors in the plant were hard to endure. It was an aromatic environment, to say the least, but designing the MX-770 was exciting work, and so no one complained, at least not much.

In June 1948 General Motors sold its shares in North American Aviation, which now became a publically owned corporation with Dutch Kindelberger as CEO and Chairman of the Board and Lee Atwood as President. As far as the engineers at MACE were concerned, it was the same superb and technically savvy leadership team at Headquarters.

Navaho I and Redstone with the 75K Engine

Bill Bollay and his Aerophysics Laboratory team were going to need a large rocket engine of at least the V-2 thrust for the MX-770 Navaho I, and they needed it sooner rather than later. Bollay's next step, and an important one, was to bring in a manager with practical engine experience. The man he selected was Samuel K. Hoffman, who for years had served as chief engineer for Lycoming, a developer and manufacturer of a line of notably reliable aircraft engines, and then became a professor of engineering at Penn State. Bollay liked that combination of industrial and academic experience. Sam was rather slight of build and not a flamboyant person, being rather quiet and normally quite serious, but his piercing gaze radiated resolve and determination. His people came to highly respect him for his quick comprehension of technical problems and his drive to be number one in rocket propulsion. Hoffman later stated, "I came out to get a rocket engine for them. Bill had a group of brilliant young fellows with no practical experience—which probably helped them with the new things that were coming along. Bill wanted me because I knew how to build engines, had built them, and brought practical experience to this young group. They all were a generation after me. I'd had a career before, but these guys hadn't, except for being in the Army and the Navy."[1] Sam set realistic goals and schedules and did not try to micromanage, giving his engineers lots of latitude to use their talents and good judgment. He was not the dynamic type of leader who gave pep talks, but he soon had earned the great respect of his team and proved to be a highly effective leader. If Lee Atwood was the most intelligent engineer in the corporation, the rocket team considered their Sam to be number two.

Starting with the German A-4 Engine

An obvious way for Hoffman to speed things up was to fully tap the German experience. That led to meetings with von Braun and his people at Fort Bliss. Dieter Huzel, a lead test engineer at Peenemunde, was hired, and he plunged right in with the design of test facilities at the new Santa Susana site. Dieter became an integral and valued member of the propulsion team. Another of

Sam Hoffman

the Germans hired for a more limited time was one of the Peenemunde Riedels, of whom there were at least three. Von Braun's Peenemunde deputy, Walter J. H. "Papa" Riedel, had gone after the war to do rocket work in England. Riedel "II," Klaus, was killed in a car crash in 1944. Riedel "III," Walther, was the one who moved from Fort Bliss to work for several years with the NAA rocket team.[2] The assessment of his contribution is mixed. On the one hand he contributed valuable details of the German development experi-

Dieter Huzel

ence, and he is credited with suggesting a rough-combustion cutoff device that saved many thrust chambers from burnout, but he was not of much help to those wanting to simplify the A-4 design, generally stating that at Peenemunde they had found features like the slow prestage startup to be absolutely essential. Rocketdyne's testing would soon disprove that.

Bollay requested and got two of the A-4 engines shipped to NAA in 1946 from the Army via its AAF contingent as government furnished equipment (GFE). These were torn down and dissected in fine detail. Then three copies of the A-4 engines were made using standard American screw threads, fasteners, O-rings, materials, and fabrication techniques. These were flow tested using water rather than propellants to confirm flow rates, pressures, efficiencies, etc., but were never hot fired. Instead a new engine was designed based on the A-4 but with improvements in obvious areas, like getting rid of the tangle of liquid oxygen lines.

75K Engine, the Americanized A-4

Before describing the redesign of the thrust chamber, we should define what exactly constitutes a thrust chamber. It is the essential device that produces the thrust for any liquid-propellant rocket engine. Its first element is the injector, which atomizes the propellants and mixes them for burning in the combustion chamber. The combustion gases are then expanded through a classic deLaval converging/diverging nozzle, with the high-pressure gas first accelerating as it is squeezed through the converging section, reaching sonic velocity at the throat and then expanding to supersonic velocity in the diverging section. The maximum gas velocity, and hence maximum thrust-producing increase in gas momentum, is achieved by expanding the gas at the nozzle exit to just match the ambient pressure. We will discuss nozzles with more sophistication in later chapters.

The bulbous shape of the combustion chamber for the A-4 made for very difficult forming and welding. Instead the chamber was redesigned by NAA engineers to a cylindrical shape of modestly reduced volume. The combustion chamber and nozzle walls were cooled regeneratively by passing the fuel through cooling passages in the thrust chamber walls before burning it in the combustion chamber. By this technique, first demonstrated in 1938 by James Wyld as a member of the amateur American Rocket Society, the heat lost to the coolant is regained (regenerated) when the fuel is burned in the combustion chamber. George Sutton was in charge of the 75K thrust chamber design and development. He adopted the conservative German feature of extra wall cooling near the injector by injecting some fuel along the walls, but George did it with a few simple orifices in the injector face rather than by drilling holes through the chamber walls.

The Germans at Peenemunde had wanted a flat plate shower head injector rather than their 18 small cups, but they did not complete testing in time to put it into production. The NAA shop machined one first out of steel and then out of much lighter aluminum, starting with a large billet and milling dozens of concentric annular passages in the injector face with alternate passages supplying fuel and oxidizer. Fuel was distributed to its annular passages through radial drilled passages bringing fuel from the thrust chamber cooling jacket. The oxidizer was distributed to its annular passages from a dome at the back of the injector body. Circular rings with drilled injector holes were then brazed onto the passages to form the flat injector face. The triplet injection pattern developed in the east parking lot testing was employed with two fuel streams impinging on each oxidizer stream. The thrust chamber looked much neater and more compact than the A-4 with only a single oxidizer line supplying the injector instead of 18 intertwined lines.

Meanwhile by 1947 the AAF had gained its independence from the Army and became officially a separate service, the United States Air Force (USAF). To avoid overlap in missile development, the USAF was given responsibility for ranges of 1000 miles and up. That meant the MX-770 Navaho I range had to be increased from 500 to 1000 miles. This dictated the abandonment of the boost/glide configuration in favor of a ramjet-powered cruise missile with a rocket boost, still integrated into a single stage. Weldon Worth, the director of the Air Force Power Plant Laboratory at Wright Field in Dayton, Ohio, took on responsibility for the development of the ramjets, contracted to Wright Aeronautical Corp., while his deputy, Col. Edward Hall, became the main pusher behind liquid rocket engine development.

Even with anticipated weight savings, the boost rocket thrust for Navaho I would have to be increased from the V-2's 56,000 lb to 75,000 lb. For the new 75K engine, it was decided to achieve this uprating by increasing the chamber pressure from 220 psia to 318 psia, keeping the throat diameter of the thrust chamber nozzle at 15.3 in., same as the A-4. The diverging section of the nozzle was retained as a conical shape of 15-deg expansion angle. To attain complete expansion with the higher chamber pressure, the nozzle area ratio (exit area divided by throat area) was increased from 3.4 to 3.6. The overall length of the thrust chamber was very nearly the same as for the A-4.

The design team was confident that with a judicious amount of fuel injection along the walls (film cooling) there would be no problem with thrust chamber regenerative cooling, but the ethyl alcohol concentration was kept at a conservative 75%. All of this resulted in improving the engine efficiency or specific impulse (thrust in pounds divided by propellant total flow rate in lb/s) from 203 to 218 s. The thrust chamber was fabricated from steel with many seam welds, so that the finished product looked rugged but would not be likely to win any beauty prizes at an art show.

The turbopump was similar in design to the A-4 but simplified to have a single one-piece shaft supported by two bearings. (The A-4 had a more complex two-piece shaft with a flexible coupling so that each half-shaft could be supported in its own pair of bearings and therefore achieve the required stiffness with a smaller diameter shaft, although at the expense of a more difficult alignment procedure.) The single-disk two-stage turbine wheel was centrally located with the oxidizer and fuel single-stage centrifugal pumps at each end.

The turbine was driven by steam from decomposed hydrogen peroxide, just as in the A-4, but instead of using liquid potassium permanganate as the decomposition catalyst, decomposition was achieved with a simpler catalyst pack of metallic screens plated with cobalt. The resulting oxygen/steam mix going to the turbine had a temperature of approximately 404°C (760°F). When this steam was exhausted, it generated another 3000 lb of axial thrust, and so the engine actually should be rated at 78,000 lb thrust rather than its more common 75K label. We will be discussing this variance in engine ratings in Chapter 5.

A pressurized helium pneumatic system was employed to activate the main propellant valves. Also borrowed from the A-4 (and Robert Goddard) were graphite jet vanes moving in the engine exhaust for thrust vector pointing to steer the missile. The final 75K engine assembly was functionally very similar to the German A-4 but much simpler in its plumbing, considerably more compact at 3.33 m (131.2 in.) tall vs 4.46 m (175.5 in.) for the A-4, and significantly lighter at 669 kg (1475 lb) vs 1127 kg (2484 lb).[3]

Wernher von Braun had been consulted during the redesign and had a continuing interest in the effort. As the design was being finalized, he arrived at the Downey plant for a visit and was briefed on the design. At that time the Aerophysics Lab had decided to use their own engine designations rather than the Air Force's XLR and LR series. The XLR43-NA-1 or 75K was the first large NAA engine, and so it would start the "A" series. They had started A-series model subheadings with the GFE V-2 engines and had gone through three redesigns, so that the 75K design shown von Braun was labeled the A-4. Wernher thought that was just great as it was the same as their Peenemunde designation for the V-2 rocket. (Actually, the final production versions of the 75K engine would be designated A-6 and A-7.) He may have decided right there that he wanted that engine for his proposed U.S. Army ballistic missile that became known as the Redstone (in honor of the Redstone Arsenal in Huntsville, Alabama, where Wernher and his German crew were assigned in 1950).

75K Testing

For development testing of the 75K engine, a large-engine test stand was completed early in 1950 at Santa Susana on the rim of a large natural bowl in the rocky hills. It provided for firing engines vertically downward and was soon identified as Vertical Test Stand #1, or VTS-1. Jim Broadston had been hired as an experienced facility manager. He placed daily operations under the management of the talented and well-liked Bill Cecka. Murray Bebbe was responsible for testing, Ed Cartotto and Phil Fons for instrumentation, Jack Dane for the machine shop, with Tom Carvey specializing in turbopump testing and Bob Lodge running the equipment lab. In addition to being an effective facility manager, well respected by the NAA top management in Inglewood, Broadston enjoyed playing the role of father figure, giving lots of advice to his young test crews on proper lifestyle, healthy nutrition, etc.

Jim Broadston

The initial firing test of the 75K engine was to be of just the thrust chamber, with propellants provided from pressurized tanks. On 2 March 1950 the observers included interested officials such as Wernher von Braun. Paul Castenholz was the development engineer for this inaugural test. As the VIPs awaited with anticipation the flame and roar of the largest rocket known to exist, the test operator started the main sequence of opening the liquid oxygen valve. The thrust chamber immediately exploded in a ball of flame—a great start. The design engineer had specified that the oxidizer dome be made of common mild steel, not being aware that it became very brittle at liquid oxygen temperatures. Under pressure it immediately shattered and ruptured plumbing near it, causing a spectacular blaze. NAA did not ordinarily look for scapegoats, but an embarrassed Dutch Kindelberger did seek out the design engineer and exclaimed, "You should have made the blasted thing out of solid gold."[4] Even precious gold would have been cheaper than the cost of the failed test. The dome was changed initially to stainless steel and then to lighter aluminum, which worked fine.

Using the German A-4 approach of starting with a prestage of about 10–15% of full thrust before progressing to full stage, the followup tests of the thrust chamber went quite well, although some of the starts were "hard," sometimes damaging the injector. After each firing the test engineer would stick his head up into the thrust chamber past the throat to examine the face of the injector for damage. This was always an interesting assignment as the resident fumes from the alcohol fuel left one with quite a buzz.

The tests were exciting, with a brilliant long supersonic exhaust jet (the carbon in the alcohol fuel glowed with great intensity in the exhaust gases) and a roar so intense it shook the body and hurt the ears even at an observing distance of over 100 yd. Only a few experienced observers, like George Sutton,[5] were smart enough to wear ear protectors—the rest played the macho game of seeing who would be the last to put his hands over his ears. Only later was potential loss of hearing recognized as a problem.

In retrospect, everything was a new experience, and the technology and facilities were rather primitive. The initial buildings at Santa Susana, including control rooms, were temporary trailers. There was no cafeteria, just another trailer that housed Goldies' food service, where a donut was a nickel and coffee just a dime. Digital computers had not yet been developed, and so the engineers had only their slide rules and mechanical adding machines. Almost nothing was known about rocket combustion processes, so that working out start sequences was a matter of trial and error. As one of Norm Reuel's rising young development engineers, Paul Castenholz, explained it, "Knowing what we did, the million and one things that could go wrong, and so forth, you approached every test with your fingers crossed. Essentially, you'd hold your ears, close your eyes [not really] and hope it would work okay."[6]

In developing a smooth start, many techniques were tested to eliminate the low-frequency chugging that frequently occurred during the low-thrust prestart. Both liquid and solid igniters were tested at locations from the nozzle exit to the injector face. A recirculating system was developed to reduce bubbles in the liquid oxygen lines, and sequences were developed to ensure smooth propellant delivery to the injector face. Lessons were even learned about the test stand. VTS-1 had been built with steel support structures built above a towering concrete base so that there would be a great distance (about 80 ft) from the rocket nozzle exit to the ground. It came as a surprise that, even with that distance to dissipate the exhaust, it was chewing away at the rock base at a great rate. A concrete deflector was tried, but it rapidly spalled away under the intense heat and pressure. Even when a steel bucket was installed at the base, the exhaust was rapidly eroding the steel. Following a suggestion by Jim Benson, a member of the test crew, a massive shower of water was injected into the exhaust, and the erosion problem was finally solved.

Combustion Problems

A few months later a complete 75K engine was installed on VTS-1, and 15 November 1950 was the historic date of the first successful main stage test of an American designed and built large liquid-propellant pump-fed rocket engine. Any celebration was short, however, as many problems remained to be solved. Techniques that had worked fine on the thrust chamber tests were producing frustrating hard starts in the full engine tests. An igniter at the nozzle exit had been fine for just the thrust chamber but no good for the full engine, which preferred its igniter at the injector face. A fuel lead sequence was superior for the chamber only but had to be abandoned in favor of oxidizer first for the engine.

Then the engineers encountered high-frequency combustion instability on the transition to main stage, which greatly increased heat transfer rates and could burn through an injector face in less than a second. The Germans had never encountered this in their A-4 engine. Years later, when combustion stability became more of a science, this was attributed to the fact that the 18 spud injectors of the A-4 essentially divided the combustion chamber into 18 small rocket motors that were much less susceptible to large amplitude acoustic modes. That understanding of the process led to the later highly successful use of baffles to break up the combustion zone into smaller areas.

In 1950, though, the problem was indeed serious and frustrating. To again quote Paul Castenholz, "You have to remember it was all relatively new. We hadn't any real experience with flat-face injectors. The Germans had done some preliminary work, but only in low thrust systems. Basically, you had to go by logic and ease your way into it without doing too much damage to the hardware."[7] On the positive side, team spirit among the test and development engineers was great. Recalls Paul Kisicki, one of the team members, "There were 50 to 60 of us and we were like family. Something would happen, and your working group would all sit down together to talk over ways of making improvements."[8]

Although the problem was new to them, the veteran Germans chipped in with ideas. Walther Riedel contributed the valuable concept of a rough combustion cutoff (RCC) device that could act faster than human monitors and thereby saved much valuable hardware. Wernher von Braun suggested and helped arrange for a small Army tank to sit near the flame bucket so that an observer could actually look up into the combustion chamber during a firing. Von Braun himself tried the tank observing post. Maddening as it was, the triplet injector had to be abandoned in favor of a like-on-like doublet pattern, with impinging pairs of fuel streams and separately impinging pairs of oxidizer streams. Although an oxidizer-lead sequence gave a smoother start, a fuel-lead sequence was better for suppressing high-frequency instability. The reasons why one was better than the other were not understood. They just went with what worked. The final successful sequence started the pumps slowly enough to avoid cavitation of the pump impellers and ran the prestage long enough to get all of the gas and bubbles out of the propellant passages and allow low-frequency chugging to smooth out so that the progression to main stage would be smooth and not trigger the destructive high-frequency "race track" acoustic mode.

That first 75K engine, known within the company as "Engine One," stayed on the test stand for six months, undergoing 65 tests, 36 at full thrust for a total of 628 s. The last 12 tests were with the same turbopump, which the post-test teardown proved to be in excellent condition. America indeed had a successful large liquid-propellant rocket engine. The only major problem was that its primary application had just disappeared.

Navaho Outgrows the 75K Engine

The confirmation that the Soviet Union had successfully tested an atomic bomb added urgency to the ultimate Air Force goal for a long-range missile—a range of 5500 n miles that could reach every important target in the Soviet Union. NAA studies were quite convincing that the ramjet-powered cruise concept of Navaho I could be evolved through the use of more sophisticated structure, lighter materials, and increased propulsion efficiency to reach that lofty goal. In the fall of 1950, just as the 75K engine was working well, NAA and the Air Force reached agreement on a new plan for the Navaho. The vehicle would be divided into two parts, with a rocket-powered booster taking the ramjet-powered cruise stage to an altitude of 40,000 ft, where the twin ramjet engines would sustain cruise at a Mach number of 2.75 or better. The full-scale mockup of the cruise stage was awesome to behold. (One needed a SECRET clearance to view it.) With a sharply pointed nose, delta wings, a vee tail, and small canard trim surfaces close to the nose, it greatly resembled some of today's most advanced fighter aircraft, but this was way back in 1950. In that age it looked like something out of a science fiction illustration. To give an indication of size, each of the twin ramjet engines had a diameter of 4 ft, big stuff in that age. In fact there is no ramjet to approach that today.

The initial goal was to achieve a range of 3600 n miles. This would require a booster with 240,000 lb of takeoff thrust. It was decided to develop a new rocket engine of advanced design with a thrust of 120,000 lb, two of which would power the booster. The new vehicle was dubbed Navaho II, or G-26 in USAF parlance. The final step would be to increase the range to 5500 n miles with Navaho III, the G-38, with a boost stage of over 400,000 lb takeoff thrust.

The 75K Powers the Redstone Missile

This could have left the 75K engine with no application. Fortunately at just that time, von Braun and his Army crew at Huntsville, Alabama, got authorization to develop a surface-to-surface ballistic missile to be called Redstone with a range of 200 miles. The 75K engine was a perfect fit, was already well along in development, and was well known to von Braun. Sam Hoffman's chief "salesman," Doug Hege, had been keeping von Braun up to date on the steady progress in the 75K development program, and so the competition for the Redstone's engine was really no contest. NAA easily won the competition with Aerojet. The Redstone required a firing duration of 110 s, and so the engine is sometimes referred to as the NAA 75-110. On 27 March 1951 a contract was signed for two prototype engines and a full development program. Modifications from the Navaho I configuration included a second peroxide tank, a larger gas generator, and a built-in ignition system. Development under Rocketdyne program manager Chan Hamlin and his successor Paul Fuller went well, and the engine was put into production.

To free the VTS-1 test stand for development work on even larger engines, all of the acceptance testing of the production 75K engines for the Redstone was moved to the Army's unused 500K test stand at the White Sands Proving Ground in New Mexico. New and apparently random incidents of destructive high-frequency combustion instability were experienced, and so the White

Hal Diem

Bill Ezell

White Sands 500K-lb thrust test stand, 1953/1954.

Sands crew, led by Hal Diem with test engineers Bill Ezell (who before joining NAA had been part of the Army test crew launching Nikes and reassembled V-2s at White Sands) and Bill Bright, was immediately back in the development business. Vital guidance to this very young crew came from the German testing veteran Dieter Huzel back in California. Combustion specialist Stan Gunn from the Research Group arrived with his high-speed Fairchild camera and photocon sensors to help study the combustion process. Combustion specialist Bob Levine from Joseph Friedman's group in Research was a valuable contributor.

The accepted start sequence was a prestart under tank head pressure before cranking up the hydrogen peroxide gas generator and spinning the turbopump up to full power and full flow. Based on some promising engine start results at Santa Susana, they tried a "full flow" start in which the peroxide valve and the main oxidizer valve were opened at once, followed shortly in just two-tenths of a second by full opening of the fuel valve. Bill Ezell sums up the result, "We were eager to try it. It was nervous time and Hal was prepared with his cutoff button, but the engine started beautifully—and so fast that the cutoff button wouldn't have helped anyway. It appeared that we had a solution to the start problem, which was subsequently borne out in an extensive series

42 Rocketdyne: Powering Humans into Space

White Sands Redstone engine test team, 1953.

Back Row: Charlie Mansur, Wayne Johnson, Bill Bright, Harvey Nelson, Ralph Vanderpool, Bill Karver, Sam Dixon, ?, Chuck Marciano
3rd Row: Margaret , Bill Baisley, C.E. Roberts, Roy Healy, Bill Ezell, Dick Hoke, Wally McHenry, Steve , ?, ?.
2nd Row: Frank Schultze, Norm Sensenbaugh, Ray Robins, Jim Stassinos, Bill Walburn, Jerry Lawhead, Ray Jolicoeur, Jim Tuck, ?.
1st Row: Mostly military

of tests in which neither 'chugging' nor high frequency combustion instability was encountered."[9]

Regular progress briefings were given to the Redstone Arsenal engineers in Huntsville. Today Huntsville is a good-sized city, largely sparked by the growth of rocket work started by the von Braun team, but in the 1950s it was a rather sleepy town without much activity. It had once been the capital of Alabama and an important agricultural center. However, it was near the northern border of the state, and the powers-that-be decided the capital should be more centrally located, and so they moved it to Montgomery, leaving Huntsville with mainly a number of impressive mansions to remind it of its days of glory. In the early 1950s it was not easy to get there—there were no commercial flights to Huntsville, which had only a primitive airport with a single narrow runway. The NAA engineers had to either fly to Memphis and catch a bus to Huntsville or else fly into Birmingham and drive for two hours north to Huntsville. Then when you got there on a hot summer day, you could not slake your thirst with a cold beer—Huntsville was "dry." The only solutions were to get invited to the Officers' Club at the Army Redstone Arsenal or else join one of several "private" clubs like the Elks Club. There soon were quite a few NAA Elks.

As the rocket engine activity grew, the propulsion group of MACE was given the name of Propulsion Center and moved in 1953 from the Downey plant to a big barnlike building on Slauson Avenue in southeast Los Angeles. It was a little bit closer to Santa Susana and had a large area for the machine shop and engine assembly, but offered nothing in the way of creature comfort improvements. On a hot day all one could do was set up fans. It was possible to open some outer doors leading to loading docks, but there was so much construction going on in the area and dump trucks roaring by that the dust was worse than the heat. Long-time Rocketdyner James McCafferty remembers, "Slauson was a factory warehouse—hot, stuffy, badly lit, with engineering and manufacturing slopping over into each other. It was close enough to the stockyards that flies were a problem. Plant engineers treated the high ceilings with fly spray, and flies died in droves, crashing down from the open rafters to the drafting tables and into typewriters."[10] Thus the design engineers sat and sweated and tried to concentrate on the tough analytical problems at hand. Finally in 1955 the rocket people got a decent new facility in Canoga Park, just 10 miles from the Santa Susana test facility. We will talk more about that later.

Redstone into Orbit

In June 1954 the Office of Naval Research (ONR) initiated a study on how to place a minimal satellite into Earth orbit. Von Braun immediately presented a Project Orbiter proposal to use the Redstone rocket as a first stage, topped with three stages of small clustered Loki solid propellant rockets. The upper stages would be spun up for stability, with no active guidance. The payload could weigh only 5 lb, which von Braun proposed to be a simple balloon about 20 in. in diameter, which could be optically tracked. Milton Rosen, manager of the Viking sounding rocket program at the Naval Research Laboratory (NRL) responded with a proposal called Vanguard that would use an Aerobee sounding rocket upper stage on a Viking booster stage. Although smaller than the Project Orbiter vehicle (takeoff thrust of only 27,000 lb vs 75,000 lb), the Vanguard had much more efficient staging and all higher performing liquid propellants, and so it promised up to 40 lb of payload.

The ensuing competition is both interesting and still somewhat controversial. For a good account the reader is referred to historian Michael Neufeld's summary in Chapter 8 of the book *Reconsidering Sputnik*.[11] The decision was made in August 1955 to go ahead with Vanguard. While there were a number of factors that influenced the selection, there is no doubt that the difference in payload was a major one.

Von Braun and his ABMA team did not give up easily. With input from the Army's Jet Propulsion Laboratory, they adopted scaled-down Sergeant rockets as better replacements for the Lokis, but that still left them shy on performance. In early 1956, soon after Rocketdyne's new Canoga Park plant opened, von Braun arrived on one of his regular visits. To a gathering of advanced design engineers and marketeers in the conference room, he explained his plan to use the Redstone missile with solid-propellant upper stages to put the world's first satellite into orbit. One hitch was that he needed about 10% more specific impulse out of the Redstone's 75K engine to achieve a significant payload, and he asked if there was any way to make that gain. John Tormey was at the briefing, and he went back to his Research group and posed the problem. One of his propellant specialists, Irv Kanarek, proposed a mix of 75% unsymmetrical dimethylhydrazine and 25% diethlyene triamine, which he was confident the engine could accept as a substitute for its ethyl alcohol without requiring modifications. In addition to increasing the specific impulse, this mix, named 'hydyne,' was more dense than alcohol, and so it could increase the propellant loading in the fuel tank and boost the thrust of the engine to 83,000 lb. Firing tests verified Kanarek's predictions, and von Braun ordered tanks of the corrosive hydyne to be mixed and shipped to Cape Canaveral. He was still not successful in convincing the powers-that-be in Washington. Thus von Braun was left frustrated as two years went by while development of the Vanguard struggled.

Then on 4 October 1957, the Soviets put Sputnik I into orbit and quickly followed it with heavier payloads, including Laika the dog. This impressive technical achievement boosted Soviet worldwide prestige and changed their image from a backward nation to a world technical leader, taking that role away from the United States. It did not help when on 6 December 1957 Vanguard attempted to put its first payload into orbit, only to fail in an embarrassing fiery explosion in full view on national television. To say the impact on national pride was devastating would be an understatement. The Eisenhower Administration had to do something quick. The Department of Defense gave von Braun's ABMA team the go-ahead. The Army crew had already designed and tested on top of the Redstone a spinning tub of 15 small Sergeant solid-propellant motors arranged to fire in three stages. It took only 90 days to integrate James Van Allen's Cosmic Ray instrument into the final-stage payload at

JPL, and on 31 January 1958 the multistaged Redstone, dubbed Juno-1 and boosted by a hydyne-fueled NAA 75K engine, put America's Explorer 1 into orbit. The weight into orbit was 31 lb—17 lb of payload and 14 lb of burned-out fourth stage. What made it sweet was that Van Allen's geiger counters on Explorer 1 and its follow-up Explorer 3 were able to detect the radiation belts around Earth, now known as the Van Allen Belts, which had gone undetected by the larger Soviet satellites.

The 75K Powers Astronauts into Space

That was not the end of the contributions of the 75K engines to the developing space race with the Soviet Union. It was announced that under Project Mercury an astronaut was going to ride a Redstone rocket on a suborbital flight. Someone at Rocketdyne had the excellent idea of putting a special tag on the engine for that flight as it progressed through the shop, reminding the workers that a human was going to bet his life on its performance. Normally the production Redstone engines were no objects of beauty. The thrust chambers were steel, and it was not felt necessary to paint over any surface films of rust. The many welds were functional but did not add to the overall appearance. However, that engine with the "manned" tag on it came out of the shop with absolutely perfect beaded welds and with the thrust chamber burnished so silvery bright it might have been chrome plated. Knowing a life was at stake made a difference. On 5 May 1961 the Redstone launch vehicle powered by that engine took astronaut Alan Shephard Jr. to an altitude of 115 miles, to be followed soon after on 21 July by a similar flight carrying Virgil "Gus" Grissom.[12] Thus the pioneering 75K engine propelled America's first astronauts into space—not a bad heritage.

Components Test Laboratory (CTL-1) Santa Susana Field Laboratory (SSFL) work shed tour jeep, 1949; back seat: Sam Hoffman, Bill Fagan, and Major R. S. Royce; front seat: O. W. Boden and Bill Cecka.

Goldies lunch room at Santa Susana, 1949.

German A-4 engine and NAA XLR43-NA-1 (Engine One) thrust chambers, 1949.

Phil Fons, engineer-in-charge of Vertical Test Stand 1 (VTS-1) instrumentation, prepares to connect the Morehouse proving ring to test fixture coupling in preparation for thrust system calibration.

MX-770 Navaho Phase 2 engine with propellant tanks, 1949.

Santa Susana Bowl Area VTS-1 construction, 1949.

Santa Susana Bowl Area VTS-1 test stand construction, 1949.

Santa Susana Bowl Area VTS-1 checkout, 1949.

First full-thrust pressure-fed thrust chamber on VTS-1. First full thrust test in March 1950.

XLR43-NA-1 booster engine (Engine One), 1950.

Santa Susana Field Laboratory (SSFL) development and large engine test crew, 1951; Back row: Pete Hummel, Bob Schuman, Larry Leas, Larry Poole, Center: Dick Agulia, Walt Hammond, unknown, unknown, Dick Werth, Warren Munson, Bill Cecka, Russ Cain, Frank Williams, Front: Norm Reuel, Jim Benson, George Jarvis, Leon Brown.

Werhner von Braun (hat) meets with Rocketdyne early rocket specialists in 1950; left to right: Walther Riedel, Tom Myers, Doug Hege, von Braun, and George Sutton; Joining them was Alfred Africano from the American Rocket Society.

Research area at Santa Susana, 1951.

First full-thrust turbopump-fed hot-fire test, VTS-1.

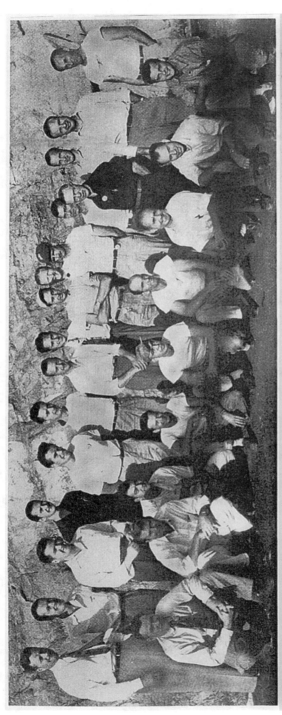

FIRST TEST CREW—Front, from left: Bill Mower, Paul Castenholz, Lyle Beach, Walt Schick, Corey Kennedy, Joe Nix, Nick Mars, Larry Leas, Joe Torma. Rear: Chan Hamlin, Hal Fowler, Bob Lumley, Larry Groeper, Bob Kilgore, Bill Leahy, Duncan Jolicoeur, Bill Cecka, Bill Nealy, Kit Doyle, Bill Yates, Warren Munson (deceased), Arthur Arnold, Harvey Nelson, Ray Keeler.

VALLEY SKYWRITER
NOVEMBER 28, 1958

Ten-year anniversary photo of first Santa Susana Field Laboratory (SSFL) test crew, 1958.

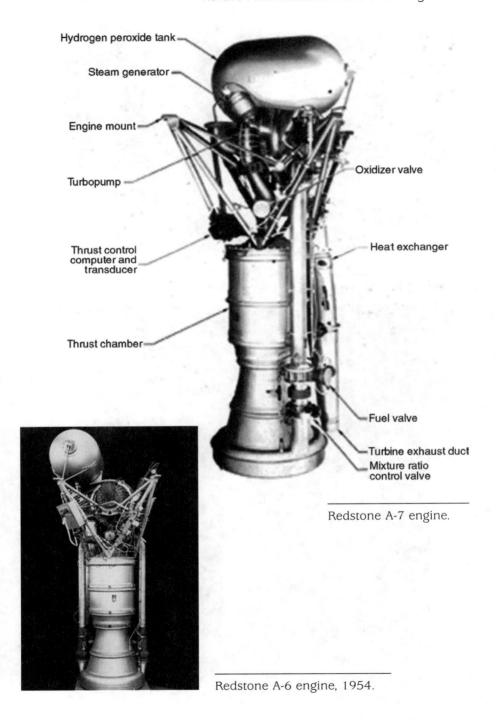

Redstone A-7 engine.

Redstone A-6 engine, 1954.

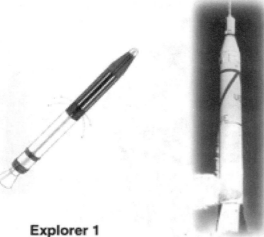

Explorer 1

Jupitor C Launch Vehicle

Jupitor C Powered by Rocketdyne Aerostone A-6 Propulsion System

Explorer 1 launch, first American satellite into Earth orbit, 1958.

Mercury-Redstone (MR-3) first manned spaceflight launch, powered by Rocketdyne A-7 engine, 1961.

CHAPTER 4

Navaho II and III and the 120K/135K Engine

To generate the 240,000 lb of thrust required to get the Navaho II missile off the ground, it was decided to use a pair of newly designed 120,000-lb engines.[1] In 1951, soon after making this decision, Bill Bollay left NAA to start his own company. Lee Atwood later admitted that he had made a mistake in having Bollay report to Larry Waite. Although Waite, an MIT-educated aerodynamicist, was technically capable, he favored rather elaborate and sophisticated management systems that did not set at all well with Bollay. Tom Dixon summarized the situation, "Bollay was oriented to university type approaches. He wasn't interested in 'management systems.' Larry Waite was an adamant fellow; he disagreed with the way Bill wanted to run things, and Bill just couldn't get along with him."[2] Bollay left just as the Aerophysics Lab that he had founded was building up a surge of new technology, capability, and contracts. As it expanded, it was reorganized under the new name of Missile and Control Equipment (MACE) Division of NAA. In their propulsion development the new 120K rocket engine would incorporate many new and advanced technical features devised by the young engineers hired by Bollay.

In the expanding rocket propulsion group, the team roles and interactions had become clearly defined. Whatever their titles, which changed as the organization evolved and grew, the enthusiastic and visionary Tom Dixon had overall responsibility for research and engineering, but it was the more conservative and detail-focused Paul Vogt who was clearly the chief engineer. Engine development under him was guided by the very bright Norm Reuel. An advanced design group to push for more advanced engine features was wisely kept out from under Vogt and reported to the more adventuresome Tom Dixon. This important group had various titles over the years, like Advanced Design, Preliminary Design and Analysis, Advanced Concepts, and Advanced Programs. For simplicity we will refer to it from here on as 'Advanced Design.' Its successive leaders (Doug Hege, George Sutton, Bob Kraemer, Sam Iacobellis, and Ed Monteath) consistently followed the policy of hiring only the best and brightest engineers they could find. This was to pay off in some dramatic advancements, like the Aerospike engine, in the years to come.

Everyone on the propulsion team settled comfortably into their respective roles, and there were really no meaningful conflicts or struggles for territory. Was it a fortuitous blend of personalities or the quiet but firm leadership of Sam Hoffman? Probably a mix of both, plus the stimulation and challenge of working with immense combustion energies and forces and trying to do what had never been done before. There was a complete focus on solving problems and a determination to make things work.

The All-New 120K Engine

The design of the new 120K engine started with the thrust chamber. Essentially the same injector as the 75K could be adapted, but the thrust chamber body for the 75K engine had been a derivative of the German A-4 and was indeed rugged but also heavy. Moreover, analysis by stress engineer Matt Ek concluded that the design could not be scaled up much beyond the 75K.[3] Matt was a big man, tall, strong, energetic, and secure in the conclusions of his own analysis. He was not arrogant, just sure of himself. Paul Vogt and the other engineers had grown to share that confidence. An alternate design approach was needed. Appropriately, an engineer at Reaction Motors, Ed Neu, had built a lightweight thrust chamber by brazing together a wire-wrapped bundle of thin-walled tubes into what was called the "spaghetti motor." It does not appear to be documented, but Hoffman's engineers had probably heard of this concept. They designed a 120K thrust chamber that started with a bundle of thin-walled 1/2-in. steel tubes that were shaped to the contour of the thrust chamber walls and then flattened into rectangular cross sections so that they fit snugly when bundled into the form of the thrust chamber. The bundle was then brazed together, and reinforcing metal hoops were added around the periphery to help contain the combustion pressure. It was fortuitous that the flattening of the tube cross section produced the smallest cross-sectional areas and hence the greatest coolant velocity at the throat where it was needed most. Later, techniques developed by a company making tapered metal golf club shafts were adapted to further tailor the tube cross sections to match exact cooling requirements.

It was known that the heat transfer rate through the thin tube walls would be much greater than through the thick walls of the 75K and A-4 chambers, and it was hoped that inefficient film cooling could be eliminated. Moreover, it was planned to increase performance by going to 92.5% ethyl alcohol instead of the 75K's 75%, but this would further increase the combustion temperature. While the heat from the 75K chamber was picked up smoothly by conduction to the flowing alcohol in the cooling passages, the greater heat transfer through the 120K's thin tube walls would produce bubbles at the surface. This would be acceptable if the bubbles rapidly collapsed in a process called 'nucleate boiling', but if the bubbles did not collapse fast enough, then the walls would be blanketed with vapor, and the tubes would instantly burn through.

There were no published test results of nucleate boiling processes with rocket propellants to confirm that it would be achieved in the new tubular

DOUBLE WALL

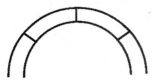

**FORMED AND WELDED
SHEET METAL**

TUBULAR WALL

**FORMED AND BRAZED
TUBE BUNDLE**

- GREATER STRENGTH
- LIGHTER WEIGHT
- MORE EFFICIENT HEAT TRANSFER

Double wall vs tubular wall thrust chamber/nozzle cross section.

The 120,000 lb sea-level thrust Navaho booster engine, 1952.

thrust chamber. However, one engineer (your author) who studied at Caltech knew that a heat transfer group at JPL had been studying high-rate heat transfer, including nucleate boiling. The problem was that JPL at that time, before the arrival of William Pickering as Director, was a sleepy research organization, and commonly let test results germinate in their files for five or more years before deciding whether to publish the data. Kraemer knew Don Bartz and Fred Gunther, the leaders of the group, and would go for a friendly visit, sitting around for an hour drinking coffee and swapping stories of hunting, skiing, and sailing (the red-headed Gunther was a madman on both the ski slopes and on the water, where he raced wild sailing catamarans) before casually asking if he could look in their files at their nucleate boiling heat transfer data. They had not specifically tested 92.5% ethyl alcohol, but one could generate a correlation that promised adequate cooling for the tubular 120K chamber with 92.5% alcohol as well as several other potential fuels. Testing of the new chamber went ahead and did indeed demonstrate very successful cooling. In fact, the cooling was so effective and tube temperatures so moderate that the brazing process was simplified to just soft silver soldering.

It is interesting that the thrust chamber throat diameter was kept at the very same 15.3 in. of both the A-4 and the 75K. Was this superstition? Probably just the hope that the new chamber would have no new acoustic modes of combustion instability. The thrust increase from 75K to 120K was achieved by raising the chamber pressure to 438 psia and the nozzle area ratio to 4.6:1. This progression in chamber pressure and area ratio would continue in the evolution of future engines to come.

Paul Vogt with Dick Ashmead and his turbopump group came up with a new design that drove the pump shaft through a geared turbine, thereby reducing the turbine size and weight while increasing its efficiency. These men were good on turbomachinery, even with the challenge of seals at cryogenic temperatures, and so pump development went smoothly. The hydrogen peroxide system to drive the turbine was eliminated by substituting a fuel-rich bipropellant gas generator. In this system, known as a 'bipropellant gas generator cycle', propellants are bled from the pump discharges to burn in a small gas generator, with the exhaust gases then passing through the turbine (or turbines) to power the pumps. The mixture in the gas generator is kept either fuel rich or oxidizer rich to keep the exhaust temperature within the acceptable limits for the turbine blades. The initial supply of propellants to the gas generator can come from several sources, the simplest being just tank head pressure, and then as the pumps begin to spin, the system bootstraps itself up to full power.

Using this gas generator cycle and a geared turbopump, the resulting 120K engine assembly was not only more compact than the 75K but also much lighter. The process to develop a smooth start and avoid high-frequency combustion instability pretty much followed the path of the 75K development—not much theory, but a pragmatic approach of test and adopt whatever works.

With the twin 120K engines for the Navaho II coming along well, attention in the preliminary design group turned toward the Navaho III. The vehicle

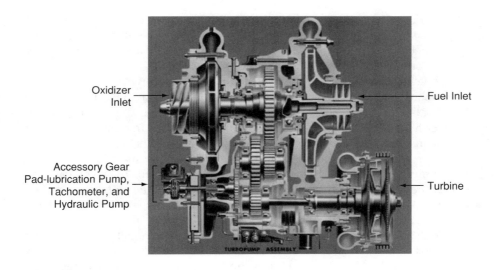

Mark 3 turbopump cutaway. The Mark 3 turbopump was the turbopump used in the Navaho G-26 and G-38; Thor MB-1 and MB-3; Delta RS-27 and 27A; Atlas B-2C, MA-1, MA-2, MA-3, MA-5, and MA-5A; and the Saturn 1 and 1B H-1 engines.

people in Downey, Dale Myers, Bob Wilson, George Jeffs, Bernie Chew and company, determined that the booster for Navaho III would require at least 400,000 lb of liftoff thrust. It was decided to use a three-engine cluster of the 120K engines boosted in thrust to 135,000 lb each. The approach to this uprating was not to perform a redesign. It was believed that there was design margin in most of the 120K engine components, and so a straightforward approach was just to run up the combustion pressure and see what broke first. The process did work. The engines with only minor changes were soon reliably generating 135,000 lb of thrust.

The next step was to integrate three of those engines into a propulsion assembly for the Navaho III booster. The thrust chambers would be hinge-mounted to provide pitch, yaw, and roll control without the drag of graphite jet vanes immersed in the rocket exhaust like those employed for the V-2 and Redstone missiles. Generating a thrust of 405,000 lb, it was by far the largest liquid-propellant rocket engine assembly in the free world. When fired, the low-frequency vibrations of the exhaust, even at a distance of 100 yd and more, would just grab and shake you. Not only would your pant legs flutter, but also your entire abdomen would vibrate like jelly. This pulsing of the stomach actually caused instant nausea in a few observers. One way or another, a 405K test was always memorable.

While the 75K was the first large NAA engine, it was really the 120K/135K design that was the parent to the next few generations of engines. Its basic features, including the bipropellant gas generator cycle, were seen in the next generations of Rocketdyne engines all the way to the F-1 for the Saturn V launch vehicle.

Navaho II made its first launch from the Eastern Test Range, Cape Canaveral, Florida, in November 1956. The autopilot failed after only 26 s of booster flight. Subsequent launches ran into failures for frequently minor reasons—one was just a failure of a minor pin attachment. Finally in January 1958 a Navaho II had an excellent boost phase and cruised for 1075 miles before one of its ramjet engines flamed out during a turn. Today you can view one of the Navaho II vehicles at the missile park near the launch sites at Cape Canaveral. It is still an impressive sight, even though its stainless steel skin is slightly wrinkled, typical of welded stainless steel sheet.

The G-38 Navaho III configuration was never built or launched, even though its three-barrel 405K propulsion system was fully developed and ready to fly. We will cover the final fate of the Navaho Program in the next chapter.

The Inspiration of Wernher von Braun

During this time period a magazine series was about to have an impact on all American rocket engineers. A recent polling (by your author) of pioneer engineers at Rocketdyne revealed somewhat surprisingly that none had been attracted to this field by reading science fiction novels. The initial attraction had been the leading-edge technical challenges of this new field of propulsion, and there was no denying the stimulation of working with such an enormous release of energy in that white-hot roaring rocket exhaust. However, to even talk about humans going to other planets was taboo—fantasy, not solid engineering. Who would want to drive across a bridge designed by an engineer who was daydreaming of space? Then in 1952 Cornelius Ryan, an associate editor at the popular *Collier's* magazine, interviewed Wernher von Braun about his views on humans going into space and convinced his editor to publish a series of articles in eight issues over two years starting in March 1952.[4] Von Braun started with his vision of a fully reusable fly-back multistage launch vehicle (the forerunner of the Space Shuttle), which would make affordable the assembly in orbit of a massive space station to permit humans in space to perform research in a zero-g environment, to conduct both civilian and military observations of Earth, and to assemble vehicles for transport to the Moon and other planets. The eighth and final article in 1954 described in great detail a mission to send a crew of humans to land and explore on the surface of Mars.

The articles were illustrated in gorgeous color by several artists, including the renowned Chesley Bonestell, whose air brush could capture the wonders of space like no other. Public response was enthusiastic, and even Walt Disney responded by featuring von Braun on television specials and having him help design his Tomorrow Land at the new Disneyland park. However, engineers were more cautious about appearing too "spacey." Even the American Rocket

Society (ARS) had dropped their original name of American Interplanetary Society for fear of appearing too "far out." After the Mars article the ARS invited von Braun to Los Angeles to address their members and answer questions about the Mars mission. The meeting in the Roger Young Auditorium was packed, and it was your author's assessment of his colleagues and fellow attendees that many had the intent to shoot down this space dreamer with technical questions. To the surprise of many, von Braun was able to give sound answers to even detailed technical questions. It was a turning point. Now one could talk about sending humans into space with rocket engines without diminishing one's credibility as an engineer. It is noteworthy that von Braun's progression of missions, from a reusable space shuttle to a space station to landing humans on Mars, is the plan that NASA adopted almost from its beginning, even though it would not appear as a formal written and approved plan for NASA until January 2004. The vehicles have been scaled down a bit from von Braun's conceptual designs, and a schedule for the Mars mission is still not set, but the plan is intact and as sound as ever.

Navaho G-26 two-engine cluster, 1955.

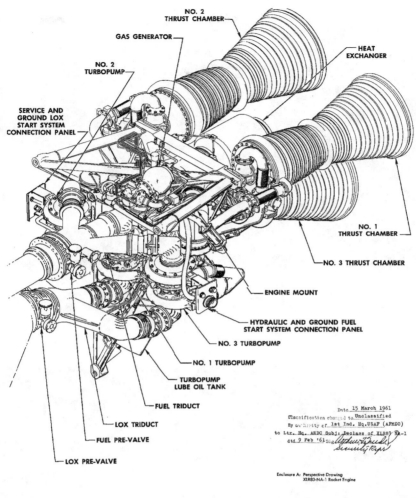

Navaho G-38 three-engine cluster perspective drawing.

Navaho G-38 three-engine cluster, 1955.

Navaho G-38 three-engine cluster propellant inlet end, 1955.

SSFL engine checkout and hot-fire control center. Observe the banks of Direct Inking Graphic Recorders (DIGR - pronounced "digger"), was a device used in SSFL test control centers to record engine and component operational pressures, temperatures, and thrust. Test data impulses activated a pen that made marks on the chart. Also Easterline-Angus (EA) Recorders were used which is a device in SSFL test control centers to record engine and component operational pneumatic and propellant valve sequencing and timing.

Back Row (L to R) Unknown, Unknown, Ken Stafford, Doug Crossland
Front Row (L to R) Ernie Barrett, Ray Keeler, Joe Harry, and Dick Werth.

SSFL Alfa Area one-year anniversary photograph.

CHAPTER 5

The Atlas with 150K and 60K Engines Orbits John Glenn

The father of the revolutionary missile that became known as 'Atlas' was Karel "Charlie" Bossart. He was born in Belgium and earned a degree in engineering at the University of Brussels in 1925. After graduate work in aeronautics at MIT, he became the lead structural engineer for aircraft designed at the Consolidated Vultee Aircraft (Convair) plant in San Diego. After World War II, because of his interest in lightweight structures, he studied the German V-2 missile design and quickly recognized the conservative nature of its structure. There was no fundamental reason why the propellant tanks had to be separate from the external skin of the vehicle. He judged that to be unnecessarily heavy. Then he reasoned that the skin need not be stiffened with structural supports—the internal tank pressure could carry the load. As an example, a child's balloon is made of thin rubber but becomes stiff when blown up.

Bossart designed a missile incorporating "balloon" propellant tanks made of tissue-paper-thin stainless steel stiffened solely by the internal tank pressure. On paper, at least, it made for an extremely light vehicle. In fact, he believed it might now be feasible to develop the ultimate strategic weapon, a guided surface-to-surface ballistic missile with a range of 5500 n miles, sufficient to reach all important targets in the Soviet Union. With 1950s technology there would be absolutely no way to shoot down such an intercontinental ballistic missile (ICBM). Great progress had been made in inertial guidance systems at both MIT and at NAA for its Navaho cruise missiles. With the accuracy of these systems, coupled with the enormous yield of the anticipated thermonuclear fusion 'hydrogen bomb' warheads, even hard targets could be destroyed. It would be a dream answer to the strategic goals of the USAF.

Charlie Bossart displays early sketch of Atlas.

Early Atlas Designs

In the fall of 1951, Bossart contacted the NAA propulsion group in Downey, already the American leader in large liquid-propellant rocket engines, to explore rocket propulsion possibilities for his new intercontinental ballistic missile. One big unknown was the weight of the as yet undeveloped H-bomb. Using the best estimate available in 1951, and even using his super lightweight balloon tanks, Bossart would need a takeoff thrust of 1,200,000 lb even if he went to the exotic propellant combination of liquid fluorine and ammonia. That thrust level was almost beyond imagination at that time, and the thought of such a large tank of the highly toxic and violently reactive liquid fluorine was enough to stagger even the most imaginative of the propulsion engineers at NAA. It may be helpful here to relate the following later experience with liquid fluorine.

In 1958 Rocketdyne accepted a contract from the USAF for the development of a complete upper stage called Nomad using liquid fluorine with hydrazine. Nomad's G-1 pressure-fed engine had an altitude thrust of 12,000 lb. By that time the hazards of fluorine were better appreciated than in the mid-1940s when small amounts were tested in the east parking lot in Inglewood. Demonstrations were performed at Santa Susana for all of the test technicians assigned to Nomad, showing them that a stream of gaseous fluorine could

Nomad G-1 Upper Stage in checkout.

Nomad G-1 Upper Stage engine hot-fire test, 1959.

readily burn a dish full of water. That made an impression. In setting up the test installation at Santa Susana, great pains were taken to achieve safety. There was an exhaust scrubber to neutralize the rocket exhaust, which was pointed upward to dissipate in the atmosphere. An expert meteorologist, Hank Weiss, and his very capable backup, Joe Glantz, were employed to assure that no test firings were conducted when there was one of Southern California's smog-trapping temperature inversion layers. Their word was final—without their approval there would be no test. The firings were generally scheduled for Sundays, when there would be a minimum work force on the Hill. The walls of the pressurized propellant tanks were made extra thick to ensure against rupture, and a liquid nitrogen cooling coil was inserted into the liquid fluorine tank to prevent boiloff and any venting into the atmosphere. Beneath the test assembly was a large catch basin to collect and neutralize any fluorine spillage. All contingencies were thought to be covered.

Then during a test a gasket at the bottom of the fluorine tank failed, leaking 600 lb of liquid fluorine. Sure enough, just as planned it was collected in the neutralizing pit. However, what had not been anticipated was that a substantial cloud of gaseous fluorine would form. Being much denser than air, this brown cloud did not readily dissipate and instead started a steady migration downhill, burning any weeds and cigarette butts in its path. Downhill meant that the cloud was heading down a canyon that led in less than 10 miles to the town of Canoga Park. The test crew was perfectly safe in its pressurized blockhouse, but what about the residents of Canoga Park? As soon as they deemed it safe, the test engineers climbed into a Jeep and started the chase down the hill after the little brown cloud. The happy ending was that the cloud did dissipate before it neared any habitation, much to the relief of all.[1] The experience did not predict a happy future for any operational missile application of liquid fluorine.

When Charlie Bossart proposed a very large missile using liquid fluorine to the team at Downey, he was advised to switch his design to liquid oxygen as the oxidizer, preferably with well-proven alcohol as the fuel. He was not especially happy about the resulting increase in the size of his ICBM design. Sam Hoffman did agree to assign one of his junior preliminary design engineers, who happened to be your author, to spend time at Convair working toward a more practical design with Bossart's team members, including Mort Rosenbaum, Charlie Ames, and a very bright young Alan Feldman (who later left Convair to became head of Aerojet's liquid-rocket programs and then of Frontier Airlines).

On 1 November 1952 America's first thermonuclear fusion device was detonated in the Marshall Islands. This was a very heavy device, not suitable for flight, but it allowed the weight estimates for a flyable H-bomb to come down into a realistic range. Then in 1953 atmospheric fallout residues from tests conducted in the Soviet Union gave positive evidence that the Soviets were also well on their way to the development of fusion warheads. Moreover, they were known to be developing large rocket-boosted missiles. There was a growing threat that they could indeed deliver hydrogen warheads to the United States. Overnight the Convair project went from an interesting conceptual design study to a very serious and urgent development program under the code name 'Atlas'. By that time in 1953, Bossart's required takeoff thrust had come down to 840,000 lb using liquid oxygen with JP-4 kerosene as the fuel. That design looked realistic and attainable, and Convair, which in May 1954 would become a division of General Dynamics, was given a go-ahead. A supporting contract went immediately to NAA. The project was given the highest priority within the Department of Defense. Not long after in September 1955, President Eisenhower elevated it still further to the highest *national* priority, second to none.

Sam Hoffman was reluctant to switch from alcohol to kerosene because it had very broad specification limits that were acceptable for jet engines but would be too variable for reliable rocket engine performance. Some batches could clog coolant passages with coke. He knew that JPL had some poor experiences with JP-4 in small motors. Under pressure from the USAF, especially from Weldon Worth, he agreed to make the effort to switch to kerosene, but with a tighter specification that produced an acceptable rocket fuel designated as RP-1. To evaluate the ease of switching from alcohol to RP-1 and to explore other performance-enhancing features, Sam Hoffman used $8 million from Independent Research and Development (IR&D) funds provided by Wright Field to set up in January 1953 a Rocket Engine Advancement Program (REAP) under one of his more senior managers, Dave Aldrich. The quick look under REAP predicted that the conversion to RP-1 was straightforward, which proved to be true. We will see that many other advancements came out of the very productive REAP effort.

Atlas Development Begins

The 840,000 lb of takeoff thrust required for the latest Bossart design could be generated with seven of the Navaho's 120K engines, five being dropped off after the initial boost phase and the other two continuing as sustainer engines in Bossart's 1-1/2 stage ICBM design. This looked like an attainable extension of the Navaho II booster, whose engines were already firing. On 14 May 1954 the Atlas Project was given full go-ahead, top 1-A priority, and a directive that "money is no object" in the path of rapid progress.

Sam Hoffman wisely picked Doug Hege to lead the Atlas rocket engine effort. Doug had that coveted M.S. degree in rocket and jet propulsion from Caltech plus valuable years of experience as a Captain in the USAF, and more importantly he had the right temperament for the job. Most engineers are quite conservative by nature—they typically are rather cautious and inclined to weigh the tradeoffs (cost, risk, etc.) before making important decisions. Tell them "money is no object" and they instinctively hold back and analyze the situation. Doug Hege, on the other hand, was more bold by nature and instantly grasped the national priority and that the imperative "money is no object" was a command for action and not just an observation. He imparted a sense of urgency to all on the program and sought out all possible means of accelerating the effort. Perhaps to moderate his own exuberance, Hege picked the cool and calm Ed Monteath as his able deputy on Atlas. Ed, like Doug, also had the valuable M.S. degree plus the professional degree in propulsion from Caltech in addition to valuable years of Navy experience, including serving as the Navy BuAer representative to Aerojet for the development of liquid-propellant rocket engines. Together, Doug and Ed made a very effective team. (In 1958 Ed would succeed Doug as program manager for the Atlas engines, and would select Jim Griffin as his assistant program manager. The very capable Bob Byron was a member of that management team.)

The drive was on to beat the USSR to the first ICBM. It was firmly believed by all involved that this was a race for national survival. That may sound paranoid to some of the younger generation today, but in the era of the tyrant Joseph Stalin and his successor Nikita Khruschev (who is remembered for taking off his shoe in 1960 and pounding it on the podium while threatening to destroy the United States.), there was a very real threat. The Atlas project was classified SECRET, and the engineers who might have access to the weight of the warhead had to get clearance to TOP SECRET and also Q-clearance from

Doug Hege   Ed Monteath   Jim Griffin   Bob Byron

The first Rocketdyne Atlas leadership team.

the Atomic Energy Commission. In visits to San Diego on Atlas, the rocket engineers had to allow a full two hours to get past the Convair security gate—it took months for their guards to learn how to deal with this unprecedented level of security. Eventually this smoothed out, Convair built a beautiful new plant a few miles outside of San Diego just for the Atlas work, and they hired a 34-year-old management whiz kid, red-headed Jim Dempsey, to direct the new division. Top-level leadership of the Atlas project passed out of the hands of its modest, tactful, good-natured, and ingenious designer, Charlie Bossart, and into the hands of more hard-driving managers.

In their primitive quarters on Slauson Avenue, the NAA rocket engineers were pressed into a maximum effort. First a 72-hour week was scheduled, 12 hours a day, six days a week with only Sundays off. A few key people were working up to 90 hours a week. It was soon found that people could not concentrate on difficult problems for 12 hours a day, especially in hot dusty conditions, without taking long coffee breaks. It was determined that just as much useful work was accomplished in a 60-hour week, 10 hours a day, six days a week.

Organizing BMD and Ramo-Wooldridge

The Air Force had assigned the ICBM program to its Air Research and Development Command (ARDC) under the direction of a hard-driving 43-year-old Brigadier General, Bernard Schriever, who had a master's degree in mechanical engineering from Stanford and was highly respected for his technical ability. Schriever set up shop in a former Catholic school house on Manchester Boulevard in Inglewood, California. Early occupants of the school house had a lot of fun with its Catholic origin, keeping the designation

"Chapel" for the largest conference room and labeling the classified files as SACRED rather than SECRET. General Schriever chose for his new organization the rather vague and innocuous title of Western Development Division (WDD), and all military personnel were directed to wear only civilian attire. WDD was later changed to the more descriptive title Ballistic Missile Division (BMD). Special paths were opened facilitating Schriever's communicating directly to the top levels of ARDC and even to the Secretary of Defense.

Development of much USAF missile technology had come from Wright Patterson Air Force Base, e.g., rocket engine technology had been advanced under the able direction of Col. Ed Hall and Bill Schnare, but a much greater depth of talent was going to be required to advance the technology required for an accurate ICBM. Schriever concluded that no single contractor had the range of talent to guide the development, and that WDD would need to assemble a systems engineering team of outstanding technical talent. To build this team, they contracted with Ramo-Wooldridge Corp. (R-W), newly formed by two top experts on guided missiles, Simon Ramo and Dean Wooldridge, who had become disenchanted with their eccentric boss, Howard Hughes, at Hughes Aircraft. R-W hired an excellent chief engineer, Alan Donovan, to guide technical development of the Atlas missile. (Later in 1958 R-W merged with Thompson Products to form Thompson-Ramo-Wooldridge, or TRW, with the more than 2000 people serving BMD in a subsidiary first called Space Technology Laboratories and then Aerospace Corp.[2])

Atlas Trims Down, Spawns IRBMs

Detailed design work had started at Slauson on the seven-engine 840K cluster when the estimated Atlas warhead weight came down again. The required takeoff thrust was reduced to 600,000 lb, which could be met by five 120K engines, four as boosters and the fifth as the sustainer. Then in December 1954 the warhead weight came down again to a firm and final weight that required only 360,000 lb of takeoff thrust. The Navaho engine uprating from 120K to 135K had gone smoothly, and engines were firing well with RP-1, and so it was decided to continue uprating development again to 150K. Two 150K engines would be the booster engines with a brand new 60K engine required as the sustainer propulsion. These engines would have to be gimbaled to provide for thrust vector control, as had already been demonstrated in a 21,000-lb thrust Reaction Motors engine flown on the Viking missile in 1949. The Rocketdyne G-38 thrust chambers had already been hinge-mounted with no difficulty and gimbal mounting would become a design feature on future Rocketdyne engines. Gimbal mounting would be a design feature on future Rocketdyne engines. Two small 1K vernier motors would provide Atlas roll control during the single-engine sustainer phase.

In January 1955 the high-level Scientific Advisory Committee, chaired by the renowned John von Neumann, pressed hard for development of a tactical intermediate-range ballistic missile (IRBM). General Schriever was concerned that it could take effort away from the more urgent ICBM development. It was finally agreed to go ahead with a 1500 n mile IRBM called Thor using subsystems spun off from the ICBM development. It would use one of Rocketdyne's gimbaled 135K engines and the two Atlas vernier thrust chambers for roll control. Schriever put it under the management of Adolph "Dolph" Thiel from the Ramo-Wooldridge staff. At Rocketdyne Tom Dixon assigned Bob Morin as project manager for the Thor engine with Ole Thorsen as his deputy and Wilbur "Willie" Wilhelm as chief project engineer. (These people went on to even bigger project management roles at Rocketdyne.)

Bob Morin   Ole Thorsen   Willie Wilhelm

The first Rocketdyne Thor leadership team.

In a complex high-level negotiation including the Department of Defense and the White House, von Braun's Army Ballistic Missile Agency (ABMA) was given a charter to develop a similar IRBM they called Jupiter for deployment by the Army and also aboard ships for the Navy. The Jupiter was in several ways a simpler vehicle than the Thor, e.g., it adapted the Rocketdyne suggestion of using the turbine exhaust for roll control rather than the more complex vernier motors. Project management for the Jupiter engine at Rocketdyne was assigned to Chan Hamlin.

Both the Thor and Jupiter IRBMs were much simpler than the Atlas ICBM, and their design and development was exceptionally fast, aided by the use of subsystems already well along in development. Rocketdyne experienced a problem of turbine blade flutter and fatigue, but that was quickly solved by putting interlocking shrouds on the blade tips. Engines were delivered within a tight schedule, and both Jupiter and then Thor had successful flights before Atlas. A decision was made in Washington to share IRBM technology with certain allies. Engineers from Rolls-Royce visited the Slauson Avenue plant and were given access to all of the detailed drawings of the Thor engines for use in the UK Blue Streak missile. A similar sharing was done with Japan's Mitsubishi Heavy Industry. The Thors went on to a long and useful life, continuing today as the Delta series of launch vehicles. The Navy (appropriately)

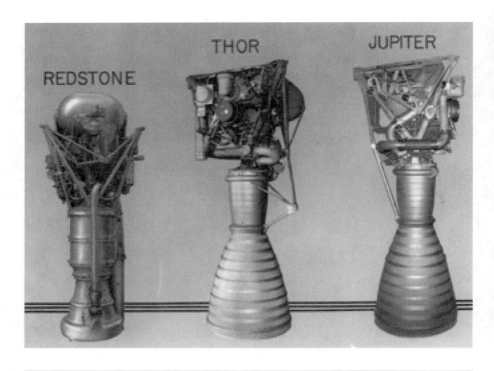

Redstone A-6, Thor MB-3, and Jupiter S-3D engines.

General Dynamics, Kerney Mesa, California, Atlas production facility.

switched from the Jupiter to the solid-propellant Polaris design and, after deployment in Turkey, further production of the Army's Jupiter was terminated as unnecessary.

To provide for production of all of the engines for these Thors and Jupiters in addition to the Atlases, the USAF gave Rocketdyne a facility in Neosho, Missouri, that was formerly operated by Aerojet. Sam Hoffman assigned the highly regarded and well-liked Joe McNamara to be manager of that plant, where he did a fine job and rose to a number of high positions, including Rocketdyne's Vice President of Liquid Propulsion Operations with all programs reporting to him. Doug Hege, along with many others, says McNamara was the best boss he ever had, and Vince Wheelock much appreciated McNamara's frequent visits to his field crews, noting with enthusiasm, "Joe Mac was a jewel and one of my favorites."[3]

Rapid Expansion

With the intense effort on the Atlas engines added to engines for the Redstone, Navaho II, Navaho III, Thor, and Jupiter, a buildup of staff was required and more space than at the Slauson Avenue plant. Tom Dixon and Doug Hege traveled to Wright Field to discuss the need for a new plant with General Irvine,

Joe McNamara, assigned as first plant manager, Rocketdyne Neosho, Missouri, facility.

the head of procurement at WAFB. General Irvine said, "Dixon, if you will build the plant, I will give you five years guaranteed production." Tom rushed back to Los Angeles to pass that good word on to Dutch Kindelberger. Dutch responded, "Dixon, we have five years of production any way you look at it. Go back and tell Irvine that the Air Force has to build the plant."[4] Dixon did, and the Air Force agreed to build the plant.

Initially the USAF planned to build the new plant at their inactive air base in Santa Ana, even farther from the Santa Susana test site than Slauson, Inglewood, or even Downey. Doug Hege recalls what happened then: "Col. Hall was at NAA on a visit. We had gone to Susie (our test site in the Santa Susana Mountains) and had returned to the valley for dinner at Helene's Restaurant. After dinner Sam, Col. Hall, and I were standing outside the restaurant, looking across Ventura Blvd. at the vacant land of the Warner Ranch, and waiting for the driver to get the car, when Col. Hall said, 'Sam, it is dumb to build the plant in Santa Ana with the test site at Susie. Why don't you buy some land here and build the plant in the San Fernando Valley?' Well we did. NAA bought and deeded 100 acres of the Warner Ranch to the Air Force and they (the USAF) built Air Force Plant #56 in Canoga Park for NAA".[5]

The new plant was ready for occupancy in the fall of 1955, and what an improvement it was. There were few offices, mainly large open "bull pen" areas, but there was good lighting and, best of all, air conditioning. The San Fernando Valley has a lovely winter climate, but it does get warm in the summer. The humidity is low, but when the thermometer gets over 105°F, it is just plain hot, dry heat or not. It was great to have room for desks and an air-conditioned building for a change.

Another major milestone occurred in 1955. On 7 November NAA officially divided the growing MACE Division of NAA into four separate and relatively independent divisions, Autonetics, Missile Development, Atomics International, and Rocketdyne, the newly adopted name for the rocket propulsion division. Sam Hoffman was named the President of Rocketdyne. When the new name was announced to employees, the reaction was not especially positive. "What's a dyne?" was frequently heard. When told it was the Greek word for "power," the discontent quieted down. The corporation executives had decreed that it had to be a one-word title, and so "rocket power" or "Rocketdyne" did indeed seem appropriate. It became a name to be proud of. It is, by the way, an officially registered trademark.

NORTH AMERICAN AVIATION, INC.
INTERNATIONAL AIRPORT • LOS ANGELES 45, CALIFORNIA

OFFICE OF THE PRESIDENT

November 4, 1955

IN REPLY REFER

TO ALL SUPERVISION:

Effective Monday, November 7, 1955, the Propulsion Center activities will become the Rocketdyne Division of North American Aviation. Mr. S. K. Hoffman has been appointed General Manager of the new division and will continue to report to Mr. L. L. Waite, Vice President in Charge of Missile and Control Equipment operations. The new division will continue to carry on research, development, and manufacture of large rocket engines and related items.

J. L. Atwood
President

NAA President's letter establishing Rocketdyne Division, 1955.

Rocketdyne Canoga Park aerial view, November 1955.

About that time a small but significant conflict occurred within NAA. The Navaho program was competing with Atlas to become the first intercontinental guided missile. Lee Atwood became aware that Convair had been given full access to the SECRET reports from the Navaho Program, while NAA was being denied any access to similar SECRET reports on Atlas. On request for equal access, he was denied on the basis of "no need to know." That hardly seemed fair. Atwood was mad. He ordered that the Rocketdyne engineer with the most familiarity with Convair's Atlas design come to give him a detailed briefing—problem. That young engineer, who had spent considerable time working with the design engineers at Convair and happened to be your author, had gone to

Convair with the agreement that he would hold proprietary the Atlas design details. He felt he could not violate that agreement. He loved his work and did not want to lose his job, but it was a matter of honor. A dilemma—preserve his integrity or his job? Sam Hoffman came to the rescue. He went to Atwood and argued that Convair (Atlas), Douglas (Thor), and ABMA (Redstone and Jupiter) were bigger customers for Rocketdyne than was NAA. If Rocketdyne was to continue to grow its business with outside customers, it would have to honor their proprietary information. The same was true for the other three divisions NAA had just created. It may not have been just or fair, but Atwood accepted that he would have to forego the Atlas briefing.

New 60K Atlas Sustainer Engine

The Atlas's totally new 60K sustainer engine design and development would have to be a crash effort to catch up with the head start on the 150K. Both engines would feature a new "bell" nozzle contour that was developed under REAP. The conical nozzle with 15-deg half-angle employed in earlier rocket nozzles had significant losses because of the nonaxial divergence of the exhaust. In other words, not all of the momentum exchange was in the desired axial direction. For a 15-deg cone angle, the divergence loss was 1.7%, which was not negligible when trying to meet the demands of an ICBM. The goal would be to have a uniform axial flow at the nozzle exit. Supersonic wind tunnels used such nozzles, as they wanted a uniform axial flow that was free of both compression (shock) and expansion waves. Before the age of digital computers, a graphic technique known as the 'method of characteristics' was employed to lay out the nozzle contour on a drafting board. It entailed projecting an array of weak expansion waves from the nozzle throat and then, where each expansion wave hit the nozzle wall, drafting a wall curvature to generate a compression wave that would just counteract the expansion wave. The resulting smoothed nozzle wall contour would produce the desired shock-free axial flow. The problem for rocket motors was that such an "ideal" nozzle was long—half again longer than a 15-deg conical nozzle.

What appeared to be promising was to turn the flow at the throat even more than for an 'ideal' nozzle, generating even stronger expansion waves, allowing sharper turning where these waves hit the nozzle wall, thus producing a contoured nozzle that was shorter than the 'ideal'. In fact, the designer could even bend the wall contour a little more at each contact point of an expansion wave, producing a net shock wave but only a very weak one, so that the losses would be small while the contoured bell nozzle would be shortened. The engines for the Atlas were designed using graphically designed bell nozzles that were about half the length of equivalent conical nozzles such as used for the Redstone and Navaho engines.

By the mid-1950s large mainframe digital computers became available, opening up new thresholds in design techniques. The Advanced Design group hired an expert aerodynamicist, G. V. R. Rao, to explore nozzle design via computer computation. Rao developed a method for determining the nozzle contour that would produce the maximum thrust for any given nozzle area ratio and length.[6] Expansion efficiency could then be traded with weight to yield an optimum bell nozzle for any given missile application. The optimum turned out to be not only more efficient but also considerably shorter by about 60% than a 15-deg conical nozzle of the same area ratio.[7] Bell nozzles today are still referred to as Rao optimum contours. While such bell-shaped nozzles would have been difficult to fabricate using the heavy-wall construction of the 75K engine, they presented no difficulty with tubular construction.

Cooling with the tubular walled thrust chambers had been so successful that chamber pressures could be raised to 719 psia for the 150K booster engines and 736 psia for the new 60K sustainer engine. The optimum nozzle expansion area ratio for the booster engines was 8:1, but the 60K sustainer would continue to fire for 368 s until high-altitude cutoff, so that its optimum area ratio was determined to be 25:1, resulting in a thrust chamber of impressive length and graceful proportions and producing an altitude specific impulse of 309 s, a new high for a production engine.

The expanding engine development at Rocketdyne was well organized, with the main design and development responsibilities clearly assigned. Under Chief Engineer Paul Vogt, Bill Brennan was responsible for directing the component group leaders, including Dick Ashmead on turbopumps, Bill Mower and Al Kramer on thrust chambers, and Perry Morris on controls.[8] Norm Reuel had the responsibility for the design and development of the overall engine systems.

Backup Role on Titan

While engine development was going quite well, there were still many technical hurdles to overcome for the complete missile system, especially in the challenging guidance subsystem, before achieving a successful ICBM. Also, there were still strong doubts about Bossart's balloon tanks. Many, like Rocketdyne's Paul Vogt, thought that they would never take the loads. General Schriever initiated a complete parallel missile development called Titan that would use more conventional tanks and with different system and subsystem contractors than Atlas. Subcontractors on one program would serve as backup second sources on the other program. For rocket propulsion Aerojet would be backup to Rocketdyne on the Atlas, while Rocketdyne would be backup to Aerojet on the Titan. This was an interesting arrangement, requiring competitors to work with one another. Rocketdyne worked with the Titan prime contractor, Martin Aircraft, as well as Aerojet in the design of a Titan propulsion system, and Hege took on another deputy, Powell Brown, to direct Rocketdyne's Titan propulsion effort. Aerojet did their own independent design—the one feature that may have come from a Rocketdyne suggestion was the use of the turbine exhaust to provide roll control for the single-engine second stage.

The Advanced Design group at Rocketdyne advocated a single engine of 300,000–400,000 lb thrust for the Titan booster stage. It would be simpler than the twin engines being designed by Aerojet and seemed like the next logical progression upward for liquid-propellant rocket engines. The new 300K/400K engine was named the E-1 and started into detailed design. About that time in early 1955, von Braun's head of vehicle preliminary design, Heinz Herman Koelle, visited Rocketdyne and was briefed on the E-1 design. He liked it and felt it would satisfy all foreseeable propulsion requirements for future large launch vehicles, even the most ambitious of von Braun's designs. In fact, he said that there would *never* be a need for a larger engine. The Advanced Design group did not argue with Herman, but clearly did not agree with his forecast.

For oxidizers in their liquid-propellant engines, Aerojet had mainly worked with storable propellants, like nitric acid. They were now being asked to develop 150K engines using liquid oxygen for the Titan I. They had to catch up to a big headstart by Rocketdyne. Just at this time the ARS had somehow arranged to have a SECRET evening tour of their members to the Rocketdyne test facility at Santa Susana. Bill Cecka, head of testing operations, organized and hosted the tour. The assured and good-looking Cecka addressed the sizeable group in the new cafeteria at Susie and confidently outlined the evening's program. He announced that they were going to witness a full-duration test of the 135K Thor engine, followed by a firing of the twin 150K Atlas booster engines, then the 60K Atlas sustainer engine, and finally the three-engine 405K propulsion system for the Navaho III booster. Now those engines were far from fully developed, occasionally had dramatic failures, and delays in testing were common. It was extremely daring of Cecka to predict such a precise schedule of firings. Bill capped off his confident introduction by saying that he had to run off to catch an airplane, but that he was sure all would go well that evening and that buses were waiting for them outside the cafeteria.

Your author happened to be the tour guide on the bus carrying Bob Young, manager of Aerojet's Liquid Rocket Plant, and his chief engineer, R. C. Stiff. The Thor engine test went for full duration right on schedule, then the impressive 300,000 lb of thrust of the two Atlas booster engines. Back on the bus there was lots of excited chatter. Then came the 60K Atlas sustainer engine, which against the dark night background had a particularly beautiful exhaust marked by luminous shock diamonds. Back in the bus I looked at Bob Young. He and his chief engineer had their heads down and were not even talking. They had to be impressed that they had a long way to go to catch up with the gang at Rocketdyne.[9] It understandably took them a while, but they did develop fine reliable engines for the Titan, and the USAF agreed to Rocketdyne's request to terminate its Titan backup effort. That also ended E-1 development.

The evening was capped off with the awesome teeth-rattling 405,000-lb thrust of the three-engine Navaho III cluster, again within minutes of Cecka's predicted schedule. It was a dramatic display of an intensity of large liquid-propellant rocket engine development that may never be seen again—a tribute to the bold support of the USAF and the accomplishments of the Rocketdyne team. Let's also commend the cool (daring) confidence of Bill Cecka.

E-1 engine hot firing at SSFL.

Navaho XG-38 three-engine cluster hot fire at SSFL.

This small example of his performance helps explain why he was one of Sam Hoffman's most trusted managers. Just like Hoffman could confidently count on Doug Hege for good marketing decisions and on Norm Reuel for dogged and intensive pursuit of solutions for engine development problems, he could always rely on Bill Cecka for not only smooth test operations, but also to represent Rocketdyne in a most favorable way with any VIP visitors to Santa Susana. Bill says that Sam treated him like a son, frequently inviting the Cecka family over for meals and the use of the Hoffman swimming pool.

Living with Exploding Growth

The growing test effort made Santa Susana, also known as 'Susie' or 'the Hill', a very busy place indeed. *Time magazine* described it nicely as follows: "*Santa Susana is a fabulous place, a three sq. mi. area fenced and guarded and crowded with up and down ridges dotted with rounded red rocks. A steep road winds over a pass and plunges into an amazing array of futuristic structures. There is no natural level land. Big buildings, fat tanks, and weird testing equipment perch on crags or nestle in rocky crannies. New construction is being pushed with frantic urgency. The whole place swarms with hard-hatted workers. Bulldozers climb like mountain goats, pushing parts of the mountains ahead of them. . . .*

Tucked away in ravines to reflect sound upward are the massive steel structures where rocket motors are put through their paces. . . . When a powerful motor is under test, an enormous flame licks downs the precipice, sometimes bounding upward in a billow of yellow fire. A sound like the rumble of doomsday rolls among the rocks, making the flesh quiver like shaken jelly."[10]

Employment at Rocketdyne to keep up with all of this development and production was soaring, to 15,000 and later to more than 20,000. The bull pen at Canoga Park was getting crowded, with desks being crammed together. The Advanced Design group, always the innovators, came up with clever modules that used minimum floor space but still gave each engineer a degree of privacy in separate cubicles with their own bookcases and telephones. This may have been the first design of the type of modular office furniture widely used today.

For years it had been the practice at NAA to hold an annual dinner meeting of all members of management where Lee Atwood and Dutch Kindelberger would give a summary of company progress and prospects. With business expanding at all four of the new divisions that had grown out of the Aerophysics Laboratory, the dinner had begun to strain the capacity of even the expansive Roger Young Auditorium in Los Angeles. Its very large main convention hall would be packed with long tables to accommodate all of the diners. Finally members of management grew so numerous that at a meeting in the late 1950s tables had to be set up in a number of smaller rooms, with the talks piped in from the main room. At that meeting Dutch moved to the podium for his talk and began in a slow solemn voice, "As I look over this impressive array of the members of management of North American Aviation, I can't help but wonder (then in a booming voice) WHO THE HELL DOES THE

REAL WORK?" That got a good laugh—members of management loved their Dutch, even if his favorites were the workers in the shop and on the assembly lines. However, the annual meeting had just grown too large and had to be discontinued.

With so many employees a method had to be developed to give people fair salary raises without taking undue time on the part of managers. Bill Guy, who was Sam Hoffman's vital right-hand man on all financial and administrative matters, assembled a team to propose an efficient solution. Bill Cecka, Doug Hege, and Norm Reuel were deeply interested in the matter of fairness in rewarding outstanding contributors and worked hard on the problem. The solution at Rocketdyne was to generate a series of salary curves with pay geared to both ability and years of experience. Each group was given a rating level, typically the 50% or "average" ability curve, and the salaries of its employees had to average out at that level, with the spread of the salaries following a statistical 'normal distribution' around that average. The managers from Advanced Design and from Research, who thought their people were the cream of the crop, successfully negotiated an average, or "centroid" 10 percentage points higher than the main engineering groups. This system required each group leader to bite the bullet and rank his people within his

Bill Guy

group's assigned normal distribution—not everybody could be number one, and some individual was going to have to be low person on the totem pole. Annual raises were then automatic without the need for a lot of debate or discussion. The system worked smoothly and efficiently and was broadly accepted as fair.

The growth of effort was also putting a strain on the test facilities at Santa Susana. There was a mammoth test stand at White Sands Proving Ground (WSPG) that was designed by the Army Corps of Engineers for engines up to 500,000 lb of thrust but that was never used. Further development (combustion instability still occurred) and acceptance testing of the production 75K engines for the Redstone missiles was moved to that 500K stand. Because the stand was mounted high on a rocky cliff, it was thought that no flame bucket would be required—wrong. The fierce engine exhausts were rapidly eroding the rock at the base of the cliff and a water-cooled flame bucket had to be installed.

In 1955 the ARS arranged for its members to observe another rocket engine firing, this time an unclassified tour to WSPG to view a 75K test. Because it was unclassified, your author took advantage of the opportunity to give his wife, Anne Kraemer, an insight into what he was working on for such long hours each week. At a safe distance bleachers were set up at the base of the cliff below the 500K stand for the observers to view the firing. A photographer took a picture of the crowd on the bleachers during the test. Everyone in the group is viewing attentively except for the one woman in the crowd, and she has her eyes closed and her hands over her ears. She now had a feeling for the exciting power of large rocket engines.

Expanding the Santa Susana Test Facility

Meanwhile back at Susie, test stands were being added at a great rate. Both the Army and Air Force managers had been pressing for more stands to accelerate testing, but they faced a tortuous path of paperwork to get adequate funds for building facilities. With a helpful suggestion from von Braun, the test stands were relabeled 'Test and Reliability Equipment' (TRE), which got rid of the problem. "Equipment" was something mounted at or on a facility, like Santa Susana. There was a more than adequate budget for TRE. In the initial bowl area, the VTS was joined by a horizontal test stand (HTS), which was initially intended to test horizontal engine starts for a plan to air-start Navaho missiles after dropping them from B-36 bombers. (USAF pilots were reported to be not overly enthusiastic about carrying a large fully-fueled liquid rocket in their airplanes.) After that plan was dropped, the HTS was immediately pressed into service in the testing of a pressure-fed engine to produce from 35K to 150K thrust for the rail-mounted Cook sleds that ran many high-speed tests at Edwards and Holloman Air Force Bases. We could technically include that engine as one that propelled humans (although not into space) as the courageous medical doctor Col. John Stapp volunteered as a guinea pig for tests of the human limits to survive high accelerations.

Horizontal Test Stand night firing at Santa Susana, San Fernando Valley in background.

Horizontal Test Stand Redstone A-6 engine hot fire.

RS-2 sled engine (160,000 lb thrust) in test at SSFL.

Cook sled at ground zero, Edwards Air Force Base, 7 August 1953.

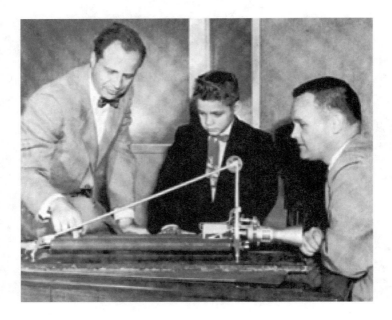

George Sutton (left) and John Tormey (right) explain test sted powered by Rocketdyne engine to young visitor Pete Clemensen (center).

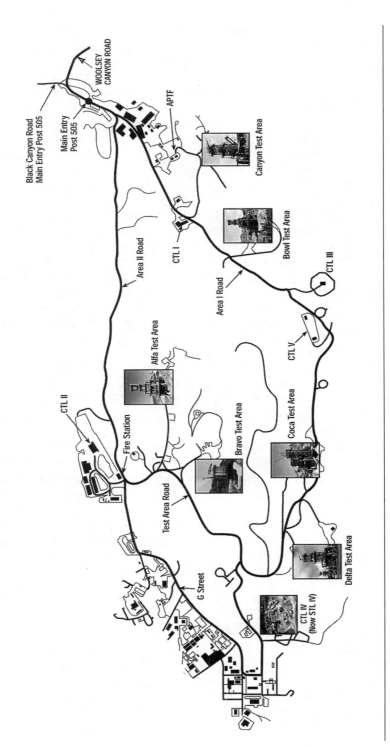

Santa Susana map of component (CTL) and engine test locations.

Additional test stands in new areas were established at Santa Susana until there were eventually three in the bowl area, three more in the Alpha Complex, three in the Bravo Complex, three in the Coca Complex, three in the Delta Complex, and another three in the Canyon Complex, for a total of 18 test stands. For further expansion, test stands were built at Edwards Air Force Base (EAFB), at Huntsville, and then at NASA's Stennis Space Center in Mississippi.

The test effort at Santa Susana was so intense that the consumption of liquid oxygen reached the level of 300 tons per day. There just were not enough cryogenic tanker trucks to climb the hill at a rate to maintain that level of consumption. A production plant was built on the site, providing not only the required liquid oxygen but also 80 tons per day of liquid nitrogen as part of the air reduction process.

With engines firing around the clock in six different bowl areas, it is amazing that there were no fatalities or even serious injuries associated with the testing of large liquid engines. The first fatality occurred when the company bought some motor scooters for the test crews to get around the facility. A technician tried to ride like a motorcycle racer, ran off the curving road, and killed himself. Bill Cecka remembers that as a continuing problem that they never completely solved. The road from the San Fernando Valley up to the test site was very curving as were the roads between the test complexes. The 1950s and 1960s were marked by a growing enthusiasm for sports cars in the United States, and their drivers could just not resist the challenge of seeing how fast they could take the curves. There were several fatal crashes.

It is interesting to note that in the experience of all crews working with large liquid-propellant rocket engines, the most hazardous element to work with is seemingly innocuous and chemically inactive gaseous nitrogen, which is commonly used to purge fumes from propellant lines and passages as well as propellant tanks. The problem comes when a technician inadvertently enters a zone of pure nitrogen. Our earthly atmosphere is composed of 78% nitrogen and only 21% oxygen, and so one experiences no new sensation when breathing pure nitrogen without the vital oxygen. Strangely, there is no sensation of gradually starving for air. Instead, without warning one almost instantly passes out. Workmen have died in empty propellant tanks that had been purged with gaseous nitrogen. Others in more open areas have been injured and even killed when falling after losing consciousness. That was the fate of two technicians at the Kennedy Space Center (KSC) who were working up in the boattail of a space shuttle. You learn to treat nitrogen with caution.

Solving Development Problems

All of the elements of the Atlas program were making good progress. The Atlas vehicles under construction at Convair were a most impressive sight. Compared to the Redstone missiles, or even the Navaho II booster, they were huge. While the tanks were being welded up out of paper-thin stainless steel sheet, they looked like wrinkled prunes rather than aerodynamic vehicles that could pierce through the atmosphere at supersonic speeds. However, once internal pressure was applied, the wrinkles smoothed out and produced a perfectly smooth skin. For transport across country the tanks were pressurized to 4 psi, and as a backup the missiles were mechanically held in tension on specially designed trailers. There was serious concern that saboteurs, or even just kids, would shoot holes in the tanks during transport. Even an air rifle could penetrate that thin skin. Surprisingly, no such incident was ever reported.

The main development problem on the rocket engines for Atlas was occasional high-frequency combustion instability in the "race track" circumferential mode, which could burn out the injector in less than a second. One apparent solution was a very fast full-flow start, which sounded like an explosion but whose chamber pressure traces were beautifully smooth. Unfortunately, Convair said the thrust buildup was so fast that it would send a shock wave up through their tanks and was more than their structure could tolerate, and so a modified start was adopted that gave good starts, but not perfect—combustion instability was still encountered between 1 and 2% of the time. That was still a concern, but well within the contractually required 85% reliability for the engines.

Great progress was made in fabrication techniques. The team of Al Kramer, Jack Hahn, and Bill Richtenberg tackled the problem of speeding up the laborious hand brazing of the tubular thrust chambers. Doing the brazing all at once in a large furnace was the solution, but getting there involved developing new brazing materials. Palladium was an important ingredient, and ways had to be found to procure it without driving up market prices. It was found that the brazing had to be performed in two steps, first with a 'thin' brazing material that would penetrate the cracks between the tubes, and then with a final layer for strength. The furnace had to be purged first with argon and then filled with a hydrogen atmosphere for the brazing process itself. The final developed technique has been used on all subsequent large engines at Rocketdyne.

Marginal lubrication of the turbopump gear train was solved by redesign of details in the lube system, gear train, and roller bearing. Fatigue failure in the turbine blades was solved by tapering the blades and by interlocking shrouds at the blade tips. These were normal development problems and not "show stoppers."

Soon the first production booster engine was ready for delivery to Convair for integration onto an Atlas vehicle. While the engine was sitting out on the Rocketdyne loading dock about to be moved onto a delivery truck, the fork lift operator swung his vehicle around and pierced the thin tubular walls of the thrust chamber—disaster. The fork lift operator was in big trouble. In no time the FBI had him in custody as a suspected saboteur. It was quite clearly an accident, and his record was perfectly clean, and so he was eventually released. However, in the urgency of the Atlas effort at that time, the "accident" was a major event. Fiberglass covers were immediately designed to protect the thrust chambers during shipment of all future Rocketdyne engines.

While in this book we will be referring to these Atlas production engines by their simple thrust levels such as 150K and 60K, you will find them referred to in documents by various other designations. For example, each Atlas booster engine was originally given the Air Force identifier of LR-89-NA-1. Then the Air Force wanted a label for the complete package that Rocketdyne was delivering to Convair, which was more than just the bare engines. It also included small pressurized tanks for starting the gas generators and for solo operation of the vernier motors, interconnects for helium and propellant lines, a relay box to accept signals from the missile, etc. The complete engine package as delivered to Convair was given the Air Force designation of MA for 'Missile Assembly A'. Thus in various refinements you will see the Atlas engine packages referred to as MA-1, MA-2, MA-3, etc. We will discuss more on that later. The engine packages delivered to Douglas for the Thor missiles were given the Air Force designation of MB for 'Missile Assembly B.'

First Atlas Launch a Successful Failure

The first flight of the Atlas was launched from Cape Canaveral on 11 June 1957. Doug Hege was there as the senior representative from Rocketdyne. He recalls the experience: "The vehicle launched from its pad with a deafening noise and a large brilliant exhaust trail from our engines—it rose into the sky. You could feel the goose pimples on your skin. We watched as the vehicle continued to climb. It reached 25,000 feet, it reached 50,000 feet, and then a number of seconds later it started to go out of control. It finally started to tumble end over end. At that point it broke up, the propellants mixed and one big fire ball exploded. Pieces were streaming in every direction. It looked like a mini atomic bomb. The first flight ended in a disaster."[11] Yet it was a very productive failure. The sight of the vehicle hurtling end over end through the atmosphere at high speed before the range safety officer hit his destruct button removed any doubts about the integrity of Charlie Bossart's balloon tanks. It was amazing that they could take loads like that. Even Rocketdyne's skeptical chief engineer, Paul Vogt, was convinced.

Certainly the USAF was convinced. That demonstration of the strength of the balloon tanks, plus excellent progress on precision guidance systems, was enough to convince them that the Atlas ICBM was indeed going to work. The next month, in July, they terminated the Navaho program. It is ironic that the spectacular advance in the technology of inertial guidance developed by the NAA. Autonetics Division was also a factor in the cancellation of the largest project at the NAA Missile Development Division. The contract termination meant that 3000 employees at the Downey plant had to be laid off. The company handled it about as well as could be done. Dutch Kindelberger personally went over the layoff list and decreed that no employee who had worked for NAA for 15 years or more, no matter his or her job, would be laid off. Some of the people on the list had skills that allowed them to be transferred to other divisions. Then tents were erected in the parking lot, and dozens of contractors were brought in to conduct job interviews with the departing employees. Testimony that respect for the company remained high is the fact that many laid off employees returned to NAA as employment openings grew in subsequent years. However, whether handled well or not, layoffs are no fun.

North American Aviation, Inc. *General Offices*

INTERNATIONAL AIRPORT, LOS ANGELES 45, CALIFORNIA

July 17, 1957

TO EMPLOYEES AFFECTED BY THE NAVAHO TERMINATION:

It is a matter of deep concern to us that the sudden termination of the NAVAHO program by the Air Force requires disruption of the team that, in our opinion, was making the greatest contribution to the nation's defense in the entire missile field.

We have been informed by the Air Force that the termination was dictated by budgetary considerations. More than $500 million has been spent on North American's portion of the NAVAHO program and related research work. As you know, the weapon system was approaching operational availability at the time of termination.

Each of you is personally familiar with the extensive research and technical achievements that accompanied the NAVAHO development. Since these were undertaken in a period when little was known of missile technology, the pioneering work by the organization also made possible many of the nation's other major missiles.

Ironically, the fortunes of defense contracting bring about a situation wherein we are forced to lay off a large number of employees at the same time that we congratulate them for a difficult job well done. It is our sincere hope that you will continue to be proud of your participation in the NAVAHO project, and that the future will bring a reversal of the present employment trend that will permit and encourage you to rejoin North American.

J. L. Atwood
President

J. H. Kindelberger
Chairman of the Board

Navaho program termination letter, 1957.

Meanwhile, after that Atlas launch failure, detailed study of all of the instrumentation and film coverage cast suspicion on the hot turbine exhaust from the engines. If the hot gases recirculated back into the missile boattail, they could have burned through the wires to the engine gimbal actuators, throwing the missile out of control. When Hege requested test time on one of the NAA wind tunnels to study the boattail flow, Lee Atwood immediately transferred a supersonic blowdown tunnel and its test crew to Rocketdyne. Tests verified the recirculation problem, which was solved for the booster engines by just canting the turbine exhausts outward into the slip stream. The solution for the centrally located sustainer engine was to inject the turbine exhaust into the main thrust chamber nozzle about a foot upstream of the nozzle exit so that the turbine gases were aspirated by the main engine exhaust. The problem was solved. As extra insurance a fireproof fabric protective cover called a boot was added to cover the nongimbaled components of the engines.

The Advanced Design group then used the wind tunnel to study the boattail flow interactions on all of its future missile/engine designs. They designed and installed a unique force balance assembly that would permit accurate force measurements in a supersonic flow while gas was flowing through the model's rocket nozzles. This capability was to become especially important in the future development of the advanced Aerospike engine design.

Production Begins

Soon Atlas launches were achieving all of their objectives. In November 1958 the Atlas went full range, a distance of 6300 miles. Three weeks later, under the closely held code name SCORE, an Atlas went all the way into orbit carrying a tape recorder with the following Christmas message from President Dwight Eisenhower, "This is the President of the United States speaking. Through the marvels of scientific advance my voice is coming to you from a satellite circling in outer space. My message is a simple one. Through this unique means I convey to you and to all mankind America's wish for peace on Earth and good will toward men everywhere."[12] It was a dramatic way to demonstrate America's technical competence and to send a message to the Soviet Union that the United States had a flying ICBM. That was an all-important and timely message to deter any planning of aggression by the Soviet leaders in the 1950s and 1960s. All of those 60-hour work weeks on the Atlas project had paid off in averting a major threat to the United States.

Production on the engines for Atlas had really cranked up, with delivery rates of complete engine packages running from 12 to 14 a month. DoD had Convair assembling missiles at a peak rate of 150 per year for test and launch and then operational deployment to 11 launch sites around the country. Then on 10 March 1960, after 46 successful flights, the 47th Atlas in the test series experienced destructive combustion instability in one of its booster engines. The Atlas fell back on its pad, erupting into a huge fireball. Could that be accepted as within the spec 85 reliability? Maybe lawyers might find it acceptable, but not the engineers. In any event, the next launch had the same failure, and so there was no debate about contractual specifications. An intense development effort followed, with input sought from any and all combustion experts, which took almost a year to find a solid solution—the use of baffles on the injector face to effectively divide the combustion chamber into a number of smaller chambers without the high-frequency racetrack mode. Meanwhile many engines had been delivered to Convair for installation on their Atlas vehicles. All of these had to be retrofitted with the new baffled injectors and the redesigned boattail boots, which was not a trivial job.

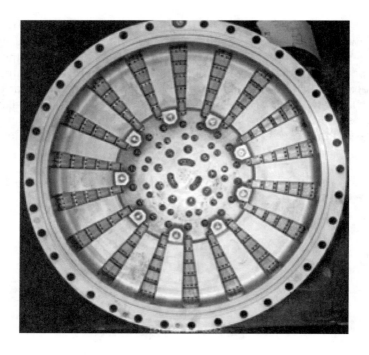

Atlas booster main injector inlet.

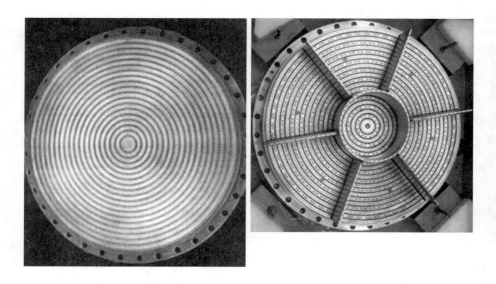

Early flat-faced doublet injector.

Atlas booster engine baffled main injector.

John Glenn Rides Atlas to Orbit

Just about the time that the newly baffled injectors were coming off the production line in early 1962, an Atlas was being prepared at Cape Canaveral to send a Mercury capsule piloted by John Glenn into orbit. Ed Monteath recalls that he vigorously urged that the booster thrust chambers on that Atlas be retrofitted with the new baffled injectors.[13] Someone (no one is volunteering for the role) decided to go with the well-tested engines as installed. In preparing for the launch, Lee Solid, lead engineer for Rocketdyne on that Mercury

Lee Solid

mission, emphasizes the meticulous attention they gave to every engine detail for that launch. Still there lingered in the minds of all of the Rocketdyne personnel their vivid memories of fiery engine explosions. The Mercury launch occurred on 20 February 1962, and Ed Monteath recalls, "They always gave us binoculars to use during the liftoff. But to tell you the truth, my hands were always shaking so badly I couldn't hold them to my eyes, so I just watched without them."[14] Even though that Mercury mission had some scary moments for Glenn during reentry (a signal indicated a loose heat shield that, if true, could have been fatal), the rocket boost phase was perfect, and so Rocketdyne propelled the first American into orbit. Rocketdyners were understandably proud on that memorable occasion. On the very next Mercury flight on 24 May 1962, an Atlas propelled Scott Carpenter into orbit, to be followed by Wally Schirra on 3 October 1962 and then Gordon Cooper on 15 May 1963.

John Glenn's Atlas booster 109D; checkout at General Dynamics/ Astronautics, San Diego plant (Rocketdyne booster engine system 115102).

Launch team for Atlas 109D, booster for America's first manned orbital flight, 20 February 1962; seated center are astronauts John Glenn and Scott Carpenter. Seated behind the astronauts is Tom O'Malley, General Dynamics test conductor, and standing in the group are Lee Solid and Bill Heer, Rocketdyne lead field engineers.

Goroon Cooper's Atlas (130D) in checkout at General Dynamics, San Diego, California.

Atlas Mercury launch vehicle being loaded aboard carrier aircraft.

Vince Wheelock, GD-A San Diego, Rocketdyne field site manager in front of an Atlas sustainer engine being processed for installation.

Rocketdyne-powered Atlas Mercury launch vehicle 107D departs General Dynamics/Astronautics, San Diego, for Cape Canaveral, with Rocketdyne field engineers Vince Wheelock and Ted Thomas (left background) joining sendoff fete. Astronaut Wally Schirra is standing with his back to Rocketdyne MA-5 engine system, 1962.

New Problems Solved

Even after engine production was going full tilt, new problems erupted and had to be solved. On 13 May 1962, less than three months after John Glenn's flight, during an Atlas F-series test at General Dynamics' Convair Sycamore Canyon test site, a sustainer engine erupted in a ball of fire. The problem was

Dick Agulia

Sycamore Canyon test stand, San Diego, California, after missile IF failure.

Atlas sustainer engine turbopump location.

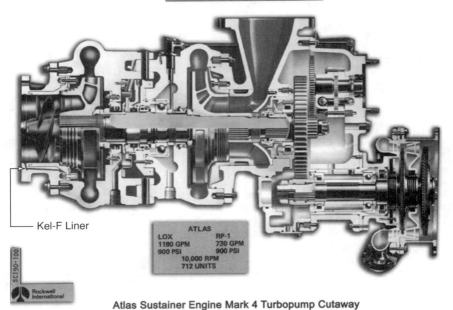

Atlas sustainer engine Mark 4 turbopump cutaway.

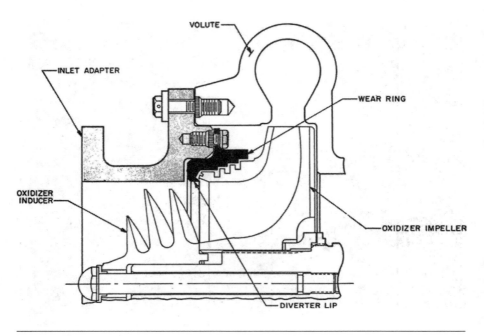

Original configuration of L0X pump and inlet adapter.

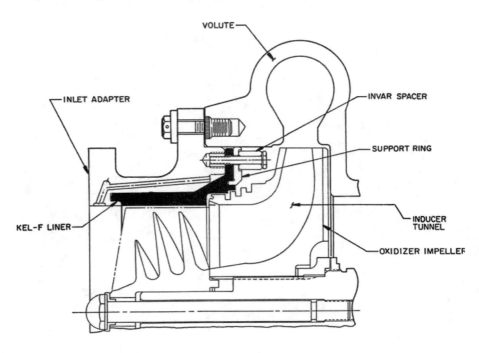

Modified L0X pump and kel-f-lined inlet adapter.

quickly traced to a rubbing of the sustainer engine oxidizer pump impeller due to shaft deflections during the high-torque pump startup. A spiraled inducer impeller had been added in the inlet to the main pump impeller to help prevent cavitation, but this inducer was at the very end of the shaft and hence received the brunt of any shaft deflections. It was found to be rubbing against the pump housing during startup. As Dick Agulia recalls, "The pump caught fire as the speed increased, which we later found was a result of the LOX pump impeller rubbing against the pump housing. The design was sound but, if the hardware wasn't perfect, you had metal touching metal and that caused a lot of friction. It wasn't something we saw a great deal of, but the isolated incident meant we had to come up with a fix."[15] The problem was quickly solved by installing a Kel-f liner, a technique still used on today's engines at Rocketdyne.

Then an Atlas was lost during a launch out of Vanderberg Air Force Base. A pressure pulse from the flame bucket during engine start was found to have enough energy to displace and partially disconnect the fireproof protective boot. The pressure pulse and subsequent inflight turbulence damaged the lube tank quick-disconnect line. This allowed lube oil to escape and the turbopump bearings to go dry and lock up. The fastening of the boot to the thrust chamber was redesigned for greater strength,[16] and Atlas performance from then on was outstanding.

In addition to its deterrent role as an ICBM, Atlas became a reliable launch vehicle for a variety of payloads, putting dozens of communications satellites into orbit and sending robot spacecraft to the Moon and to all of the planets from Mercury to Saturn. The 1-1/2 stage Atlas, a real performance "hot rod," stayed in production for an amazing 47 years, achieving a record of 77 consecutive successes.

Engine Designations

It is perhaps a bit puzzling to follow the MA designations for the complete three-engine packages delivered to Convair for the Atlas vehicles. The early MA-1 R&D version integrated three fairly independent engines. For simplification in the production MA-2 assembly, the two booster engine turbopumps were mounted together and attached to the missile structure so that they could be powered by just one gas generator instead of two. Later, at USAF request to simplify logistics and allow ready substitution of engines, an MA-3 version was produced that put the turbopump and gas generator back on each booster engine so that the engines were identical and interchangeable. For space launches the simplicity of the MA-2 configuration was preferred, and an upgraded engine package labeled the MA-5 was delivered with each booster engine now clear up to 215,000 lb of thrust at takeoff.[17] (MA-4 was the proposed single chamber E-1 engine for the Titan. It never found an application and was not fully developed.)

In this book we have generally avoided the Air Force engine designations as being more confusing than helpful to the general reader. The Air Force LR engine designations seem a bit complex, and they were not widely used within the company. Similarly, the MA and MB numbers do not directly tell anything about the engine. The company did initiate an alphabetical series, starting with its first production engine, the 75K engine for the Redstone, as its 'A' series. None of the engines for the Navaho missile went into full production, and so they did not get an alphabetical designation. When the Atlas engines were ordered into production, Sam Hoffman charged Ed Monteath with coming up with a continuing numbering system for Atlas and all future engines. Monteath logically proposed continuing the alphabetical sequence, and so the next production engine would have a 'B' designation, the following one would be 'C', and so on.

If the Monteath plan had been followed, the next production engines, for the Atlas booster, would have been a B series and when packaged with the Atlas sustainer engine would be designated C. The engine for the Thor would be a D. However, none of those B, C, D designations were ever used. The Monteath alphabetical sequence did pick up again on the 'E-1' engine and beyond, even though the E-1 never made it into production. The following 'F-1' became famous as it boosted the Apollo astronauts to the Moon. 'G-1' was a daring fluorine/hydrazine propulsion system, while 'H-1' was the engine that powered the Saturn I and IB boosters. The letter 'I' was too much like a 'one,' and so it was skipped, leading to the very successful 'J-2' upper-stage engine.

Why were the B, C, D designations skipped over? Monteath says that Sam Hoffman really wanted a designation system that reflected the power of the engines.[18] When he was chief engineer at Lycoming, developing small aircraft engines, his dream was to build really big engines like Pratt & Whitney. At Pratt & Whitney they showcased the might of their piston engines by designating them by engine displacement. For example, the designation for the powerful P&W R-4360 engine advertised its huge displacement of 4360 cubic inches. Hoffman wanted the designations of Rocketdyne engines to reflect their impressive thrust levels. Thus Sam preferred calling the Redstone engine the 75K rather than the A-7, and so 75K was what it was most commonly called within the company. For that reason the engines following the 75K were known in-house as the 120K, 135K, 150K, and 60K engines. In this book we have attempted to simplify engine identifications by sticking with thrust-level designations until the point where alphabetical designations were clearly accepted in-house, starting with the E-1.

Unfortunately, even the labeling by thrust levels is not all that simple. Sometimes the thrust of the turbine exhaust gases is included and sometimes not. Hence the thrust of the A-7 is sometimes listed as 75,000 lb and sometimes 78,000 lb. Within the company the engines were most commonly identified by the thrust level in pounds (never the metric 'newtons') of the main thrust chamber only, without the contribution of the turbine exhaust. Therefore, the A-7 engine was really known as the 75K engine (not 78K), the engine for the Navaho II was the 120K, its uprated version for the Atlas booster was the 150K, etc., rather than by their total net thrust output. In this book we have mainly referred to those main-chamber thrust levels. As an example, we identify the cluster of three 135K engines for the Navaho III as the 405K assembly, but one will also see it identified in some references by its 415K net thrust. Is this a bit confusing? Your author is afraid so—sorry about that.

The evolution of large engines at Rocketdyne will be identified in this book as 75K, 120K, 135K, 150K, 60K, E-1, F-1, G-1, H-1, J-2, X-1, SSME, and then the RS series in current use. This may not look consistent, but that is the way it was within the company.

Santa Susana staff meeting, 1954.

Atlas liftoff showing two Rocketdyne booster engines (outside), one sustainer engine (center) and one of two verniers (upper center).

Atlas Mercury launch, 1962.

CHAPTER 6

H-1 Powered Saturn IB Orbits Apollo 7, Skylab Crews, and Apollo Soyuz

This chapter highlights the persistent drive of Rocketdyne engineers toward engine simplification. We will start with an engine that should have been built but never quite came to be.

A Simple Engine Design for Manned Flight

Because the focus of this book is Rocketdyne's contribution to boosting humans into space, it is appropriate that we take a quick look at the rocket-powered X-15 aircraft, which in 1963 reached an altitude of 354,200 ft (67 miles) that exceeds the generally accepted 50-mile lower boundary of "space." Certainly the test pilots that flew the X-15 had some right to their claim that they were among the first astronauts. In the end the X-15 was not powered by a Rocketdyne engine, but we will maintain that it should have been.

NAA started the design of that remarkable rocket-powered plane early in 1955. Heavily involved was test pilot Scott Crossfield, who had been battling Chuck Yaeger at Edwards Air Force Base (EAFB) in their personal race to set records in a series of experimental supersonic aircraft. Scott knew the X-15 would shatter all of the records for speed and altitude, and he wanted to be the first to fly it and set all of those records, and so he left Edwards and went to work for NAA in Inglewood and played a leading role on the X-15 design team.

The NAA X-15 designers in Inglewood requested help from the NAA rocket propulsion group at the Slauson plant. Rocket engine explosions were still pretty common in those days, and the Advanced Design group was especially concerned about safety. They already had a favored approach for a piloted aircraft engine. The NAA aircraft designers at their Columbus, Ohio, plant had wanted to extend the operational usefulness of the famous F-86 Saberjet fighter and its Navy version, the FJ-4, by adding a small super-performance rocket engine to boost it to supersonic speeds. The Advanced Design engineers decided the best approach was to follow the German lead on their Helmuth Walter Company engine for the ME-163 interceptor using hydrogen peroxide

North American aircraft X-15, the winged rocket record breaker and space prober.

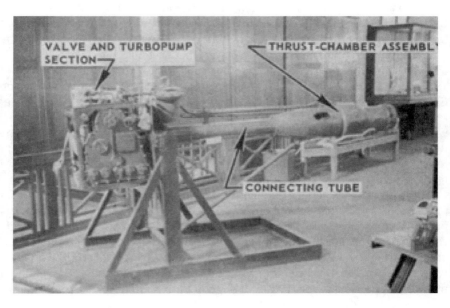

Walter's rocket engine for Me-163 airplane interceptor.

as the oxidizer. The beauty of using peroxide was that its monopropellant properties could permit a beautifully simple propulsion system. The resulting hydrogen peroxide/JP-4 jet fuel AR-1 engine and its later AR-2 and AR-3 upgrades were models of simplicity. The pilot had a throttle lever directly connected to a valve that controlled the flow of peroxide to a monopropellant gas generator whose catalyst-decomposed hot gas started to spin the turbopump as the valve was opened. As the peroxide pressure built up, it directly opened the main oxidizer valve and admitted peroxide into the catalyst pack in the main combustion chamber. The resulting buildup of steam pressure opened the jet fuel valve with resulting spontaneous combustion as the jet fuel hit the hot oxygen/steam mix. The 6600 lb of thrust could be readily throttled over a range of 50% and more, and there was no problem with restarts. Most importantly, the system design eliminated just about any failure mode that would allow unburned propellant to accumulate in the combustion chamber and cause an explosion. The only time when fuel could enter the combustion chamber was when hot steam/oxygen was present, guaranteeing that the fuel would always immediately burn. There was no electrical system or pneumatic/hyraulic system required to actuate the valves, just the one direct connection of the throttle lever to the gas generator valve.

AR2-3 Aircraft rocket booster engine.

Development testing of the AR-1 engine progressed smoothly, and the engine was soon installed in an FJ-4 fighter airplane at the NAA plant in Columbus, Ohio. The film of the first in-plane firing test, complete with pilot in the cockpit, is interesting. As the countdown approached zero, the Columbus engineers all crowded closer to the airplane. The film shows that a few engineers are backing away from the plane—all of those were from Rocketdyne. They had to know how fail-safe the engine was, but that knowledge could not overcome years of prior experience with exploding rocket engines.

The AR engines worked like a charm with no explosive "incidents." They flew 103 flights in the FJ-4 and another 31 flights in an F-86, reaching a level flight speed of Mach 1.31. While the peroxide supply lasted, the planes could consistently win dogfights with the most modern supersonic fighters of the day. The only complaint from the pilots was that, when the peroxide was exhausted and the engine shut down, the reduction in speed was so abrupt that they felt like they were being propelled out through the windshield. The engines were later mounted in more modern F-104 aircraft for an additional 302 flights, carrying those fighters to an altitude of 121,000 ft.[1]

When Crossfield and company came to the Rocketdyne engineers, they proposed that this system just be scaled up to the 60,000-lb thrust level required for the X-15. With this engine it would have been Rocketdyne that propelled the X-15 into space. Instead the decision was made by managers in the Air Force and NACA to adapt the Reaction Motors oxygen/anhydrous ammonia XLR30 engine already under development by BuAer for the Navy's Viking rocket vehicle. The reasoning was that it could be adapted sooner and at lower cost than any new engine. That did not turn out to be true.

From the beginning safety was a problem. It was essential to avoid any accumulation of explosive unburned propellant in the combustion chamber. Reaction Motors engineers finally ended up with a design incorporating three stages of ignition, starting with a spark-plug-ignited stage in a small chamber whose hot gases then ignited an intermediate stage chamber that in turn ignited the full stage—essentially three combustion chambers in series. Thrust chamber cooling also proved to be difficult. As the schedule slipped and costs escalated, Rocketdyne was contracted to modify its Atlas 60K sustainer engine for incorporation into the RMI engine, now designated the XLR99. It soon became apparent that major modifications would be required to adapt the Atlas engine to the three stages of ignition for the XLR99 and that backup effort was dropped.[2]

In one of the early ground tests of the XLR99 engine installed in the X-15, Scott Crossfield was in the cockpit with the canopy lowered and locked in place. On start the engine exploded. With a fire raging behind him Crossfield, probably the coolest test pilot of all time with no visible signs of ever having a nervous system or adrenal glands, remained comfortable in the cockpit knowing that the X-15 had been designed with massive heat protection. However, his technician thought he was in danger and rushed to the plane and started struggling to open the canopy. Crossfield could see he was badly burning his hands, and so he reluctantly opened the canopy and rushed the man away.

The engine almost got him again during a flight test. After the X-15 had been dropped from the mother B-52 aircraft, its engine failed to ignite. Crossfield started dumping propellants, but there was too little time as the aircraft plunged toward the desert floor. NAA later made a film with sound track to document the flight. It shows the X-15 nose high but dropping like a stone, almost vertically. On impact it breaks nearly in two, with clouds of vapor but no explosion. The camera zooms in to Crossfield leisurely unbuckling his harness, climbing out of the cockpit, and calmly strolling away from the steaming wreck. A crash truck roars up, and the crew tries to get Crossfield aboard to rush him to the hospital. Instead he brushes them aside, picks up a tape recorder, and proceeds to dictate his flight report in an absolutely calm monotone. The guy just had no nerves—the ideal test pilot. With the RMI engines the X-15 went on to set speed and altitude records that still stand today. Your author is no doubt prejudiced, but believes that with the far-simpler and fail-safe Rocketdyne peroxide engine, the X-15 and Crossfield would not have had so many memorable close calls.

Working with the Air Force

The engineers at the USAF Propulsion Laboratory at Wright Field were great to deal with in their support of engine simplification as well as advances in performance. It was always a pleasure to work with them, even though getting to meetings with them in the 1940s and 1950s was not so pleasant. To get a better feeling for what it was like to do business as an aerospace engineer in those days, let us take a brief look into travel conditions.

Flying in a noisy Douglas DC-4 to Chicago left one half deaf, and then came a very slow flight in an aging DC-3 to Dayton. At Wright Field they had no on-base bus system, and so in winter one had to slog through the snow to the propulsion building dragging a heavy padlocked 3×4-ft aluminum case carrying the large cardboard poster boards that were used to give briefings in those days. It is pretty hard today to imagine such crude briefing materials.

Later on better airplanes came on line. The DC-6s were faster even if not quieter. The main competitor for these Douglas airplanes was the very graceful and curvaceous Lockheed Constellation with its distinctive triple vertical tails. It was a bit quieter and more pleasant to fly in. Then came the turbocharged Douglas DC-7s that could fly nonstop from Los Angeles to Washington, D.C., in just seven hours. Again, these airplanes were not at all quiet, but they made trips to the Pentagon shorter and hence more endurable, especially if one took an overnight flight and could find open seats to lie down and sleep the entire trip. At the same time Howard Hughes had purchased a number of new Constellations for his Trans World Airlines (TWA). The new model, called the Jet Stream, had a longer wingspan enabling it to fly higher and hopefully smoother. Unfortunately, it took eight hours to fly across country, and business travelers were opting for the faster DC-7s.

To get back in the competition, TWA turned the Jet Streams into luxury flights. They featured full recliner seats with elevated leg rests. After takeoff the cute young flight attendants (who were not allowed to be married in those days) took off your shoes and gave you cloth slippers to wear. Then came a couple of rounds of cocktails. Dinner started with a white wine followed by a nice red wine with the main course. Champagne of course came with dessert, and then a fine after-dinner liqueur. That was enough alcohol to facilitate one to sleep the rest of the way across country, but the flight attendants would then come down the aisle asking cheerfully, "Cocktails anyone?" It was almost beyond belief, but some of the onboard salesmen would take them up on the offer. On deplaning, TWA would have a group of husky men lined up to help a number of wobbly passengers off the plane. Because it left Washington for the trip back to Los Angeles at 1:00 p.m., the flight was widely known as "The Noon Saloon." TWA stayed competitive with these luxury flights for several years until the much faster and quieter jet-powered Boeing 707 swept past all of its competitors.

The Role of Advanced Design

At Rocketdyne the drive of the Advanced Design group for simplicity was not limited to engines for aircraft like the X-15. They believed that many of the features adopted from the German A-4 engines could be eliminated, even though Walther Riedel kept advising them otherwise. The 120K engine had made a major step forward when it replaced the entire hydrogen peroxide gas generator system, including its separate pressurized peroxide tank and plumbing, with a bipropellant gas generator fed from the main turbopump. A deterrent to further simplifications was the chief engineer, Paul Vogt, who was brilliantly persistent on design details but resisted trying new features. His judgment was that one should never try more than one new engine feature at a time. That was a very safe approach but greatly slowed any drive to simpler and more advanced engines.

Vogt was a very interesting man. He was sturdily built, increasingly bald, and with piercing eyes that would bore into you through his Ben Franklin eye glasses. He was very sharp technically, had strong opinions, and was a tough opponent to debate. He had a great capacity for detail, which made him a strong contributor in solving tough engine design and development problems. He was highly respected by his design and development engineers and, when not debating a design feature, he could actually be very personable. Unfortunately, his ultra-conservatism made him the kiss of death for any new advanced feature being proposed by the Advanced Design group.

Key at this juncture was the vital role played by Tom Dixon, who had various titles over the years but always functioned as the principal engineering manager pushing for technology advancement. In that role he always outranked Paul Vogt. Sam Hoffman was the solid and determined leader and organizer, but for technical advancements he depended heavily on Tom's judgment. Two of their joint organizational decisions were of special importance. First was the establishing of an Advanced Design group that was maintained independent from the main engineering groups directed by Vogt.

Initially under the leadership of Doug Hege and George Sutton, there were no bounds put on Advanced Design's pursuit of advanced rocket features. We will be covering many of them, like engine simplification and the progression in performance to the very advanced Aerospike engine. There were many early members that led the innovative work in Advanced Design, including Phil Albanese, Vern Degner, Bob Dillaway, Vern Larson, Al Sutor, Hank Wieseneck, Jack Conyers, and Mark Hoffman, to name just a few. Over 20 years leadership of the group passed from Doug Hege to George Sutton to Bob Kraemer to Sam Iacobellis to Ed Monteath.

It is interesting to note that their pursuit of advanced propulsion was not restricted to large liquid-propellant rocket engines. A young engineer, Marty Willinski, was full of new ideas and in the 1950s was one of the early proponents of ion propulsion, which is now being used for very high efficiency propulsion on spacecraft.

Doug Hege

George Sutton

Bob Kraemer

Sam Iacobellis

Ed Monteath

Over 20 years of leadership of the Rocketdyne Advanced Design group represented above.

Quite an extensive effort in Advanced Design was devoted to a joint effort with the Marquardt Corporation on a liquid air cycle engine (LACE) that would act in an airbreathing mode as it accelerated through the atmosphere, at the same time condensing liquid oxygen out of the air, and then use that liquid oxygen in a final rocket mode out of the atmosphere and into space. Advanced Design was pushing for NAA to purchase Marquardt for the further pursuit of that promising concept. Instead, the originally SECRET idea was eventually declassified and went into the storage bin, to be resurrected almost 50 years later as NASA pursued more efficient ways to launch payloads into orbit.

The Advanced Design group also pursued advanced development on solid-propellant rockets, although Rocketdyne was during the 1940s and early 1950s in and then out of the solid-propellant business after starting a pilot production line that produced several thousand NAKA 1.5-in. air-to-air rockets, and then in and out again with their Astrodyne plant in MacGregor, Texas.[3] In the Advanced Design group, Charlie Morse, with support from Randy Sheeline, was a key idea man on solids. In the early 1950s the size of large solid motors was limited by having to cast the propellant in the motor casings one batch at a time. Mixing large batches was a dangerous process that not infrequently resulted in major explosions. Morse suggested bringing the semifluid propellant ingredients together in a small mixing pump, so that the hazardous mixing would involve only a small amount of propellant. The entire propellant grain would be cast in one continuous process. It was tried and worked, but not further pursued by Rocketdyne or Astrodyne, as the company did not choose to enter the large-solid business.

Morse also pondered the fact that powdered aluminum was and is commonly added to the solid-propellant mix to increase performance. Morse reasoned that if the aluminum were incorporated as fine wires instead of powder, it could contribute structural strength and lighten the casing. The grain would be designed for a decreasing chamber pressure so that as the wire was consumed, the pressure load on the casing would decrease. This was clever thinking, but that work was stopped when Astrodyne was sold.

The Experimental X-1 Engine Pioneers Simplicity

A second very important organizational move instigated by Tom Dixon and Sam Hoffman in 1957 was the establishment of an Experimental Engines group and giving it a free hand to pursue the testing of advanced engine features at their own discretion without any chain of formal approvals. While Sam Hoffman, Paul Vogt, and the main body of Rocketdyne engineers concentrated on large engine development with great determination and persistence and produced engines of enormous power and excellent reliability, one must give Tom Dixon a great deal of credit for establishing this small but vital experimental group and for picking Paul Castenholz to head it. Paul was the ideal choice for the job. He had the desired experience as both a test engineer and an engine development engineer. More importantly he knew how to shortcut the system and get experimental hardware built and tested in a hurry. With encouragement from Tom Dixon, he was off and running, with a charter to try bold new engine features at the group's own discretion—no higher approvals were required. As Paul puts it, "We had no restrictions, no holds barred."[4]

Castenholz assembled an excellent team of "do'ers." Joe Vehige was the team's chief designer, with support from specialists like Dick Francis on pyrotechnics. Key members of the development team included Bill Ezell, Jim Bates, Cliff Hauenstein, Max "Mike" Yost, and the up-and-coming Dominic "Dom" Sanchini. One of the test engineers hired for the group by Bill Ezell was Dick Schwartz, a very bright and hardworking ('workaholic?') young man who by 1983 was to rise to the role of President of Rocketdyne.[5] Of course it would take money to try new things, and here Col. Hall and Bill Schnare and their crew at the WAFB Propulsion Laboratory were strong in their support. Between a generous IR&D allocation and Atlas Product Improvement funding, the total budget at Rocketdyne for exploring advanced engine features reached $25 million per year. Through the management channel of REAP, the Experimental Engines effort was amply supported.

The early work of the group was research and technology oriented, following Castenholz's early work under John Tormey. Different combustion chamber shapes were tested, with windows in the chamber wall to view the combustion process. High-speed pressure traces were recorded to study any combustion instability. Different ignition sources were tested, including pyrotechnic devices mounted to the injector (rejected as causing damage to any jet vanes or to future tubular-walled thrust chambers) and von Braun's suggested igniters that swung out of the nozzle after ignition. The final choice was a starting slug of hypergolic propellants that ignited spontaneously when

Dick Schwartz

injected ahead of the main propellants. This method was adopted for future large engines. Alternate thrust vector control schemes to replace jet vanes were tested and evaluated, including the injection of fuel into the main exhaust stream in the nozzle. That never worked with reasonable efficiency. Alternate thrust chamber gimbaling arrangements were tested, some with flexible lines and some with multiple rotating joints. Both worked. Then the focus turned to engine systems.

Paul Castenholz had a reputation of being a "lone wolf," not given to being a team player, but the Advanced Design group under Bob Kraemer found him very receptive to suggestions, and regular coordination discussions were held with Paul. A primary goal was established of simplifying the engine systems. The first step was to get rid of the complexity of a prestart sequence. That was easy, as both the Redstone and the Atlas engine development had already demonstrated smooth full-flow starts, and Experimental Engines had the basic 150K Atlas booster engine as a starting point. The use of a bootstrapping bipropellant gas generator had already proved feasible for the Atlas engines, but Atlas still required pressurized start tanks to start the gas generator up to a bootstrap level. Thus the next step in the evolving experimental engine, now named the X-1, was to replace the pressurized start tank system with a simple solid-propellant cartridge to start the turbine spinning. This worked like a charm once it had been determined how big a charge was required to get the pumps to the point where the bipropellant gas generator could take over and bootstrap the system up to full power.

The main propellant valves had always been actuated by an electrically controlled pneumatic system. Why not eliminate all of that by having the RP-1 fuel serve as the actuating fluid? That way as the turbopump spun up to speed, the buildup of fuel pressure would directly actuate both the fuel and oxidizer main valves. This became identified as a 'pressure ladder sequence' and greatly simplified the engine controls—the entire control system for engine starts was essentially just the two wires that ignited the solid-propellant squib, which started the turbopump turning. As pump discharge pressures built up, the propellants were fed to the gas generator, and the system bootstrapped its way to full power. The buildup of fuel pressure opened the main propellant valves while forcing a small hypergolic igniter charge into the chamber, and the engine was off and running. It worked beautifully. Gone was the electrical sequencing system, the high-pressure pneumatic system, and the start tanks and associated plumbing—terrific.

Engine changes were coming so fast that there was danger of confusion among the test engineers. Mike Yost, a key member of the test team, recalls, "Paul Castenholz, the Experimental Engines Group Leader, insisted that each of us carry with us at all times a copy of the current system schematic, and sometimes challenged us regarding design issues by asking to see our personal schematic to assure that we were speaking for the most up to date design configuration."[6] When you worked for Paul, you did things his way, or else you were off the team, but this procedure was accepted as a very good discipline.

H-1 Powered Saturn IB Orbits Apollo 7, Skylab Crews, and Apollo Soyuz

The Thor propulsion system had adapted two of the Atlas 1000-lb thrust vernier motors for roll control. The X-1 was used to demonstrate that these could be eliminated by running the turbine exhaust through a hinged nozzle to provide control forces in roll. This feature immediately went into production for the Army Jupiter IRBM. Other simplifications were in the details, e.g., the turbopump oil lubrication system was simplified to using just the RP-1 fuel with the injection of a small amount of lubricant additive.

A wilder idea was to try to eliminate the gas generator entirely by tapping gases off a fuel-rich zone of the main combustion chamber to drive the turbopump. Starting would require a substantial source of energy, and stabilizing and controlling the thrust level would require a high-tech hot gas valve. Why not give it a try? The X-1 was so modified in a configuration identified as the X-4, and to the surprise of many, it worked. A few years later the tap-off power cycle was successfully demonstrated again in an experimental hydrogen/oxygen engine labeled the J-2X. However, getting the right temperature hot gases in the tap-off power cycle was difficult, and thrust control stability was marginal, and so the direct tap-off feature was never really pushed for production. The other simplifications did go into production in an engine we will discuss called the H-1.

X-1 engine.

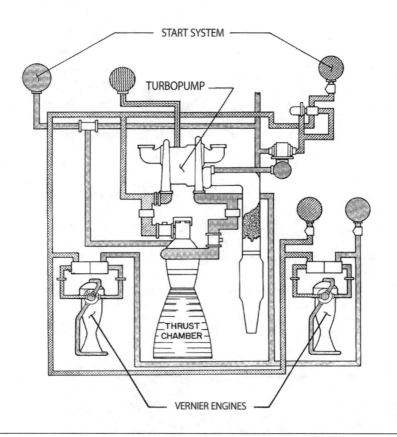

Thor MB-3 Schematic.

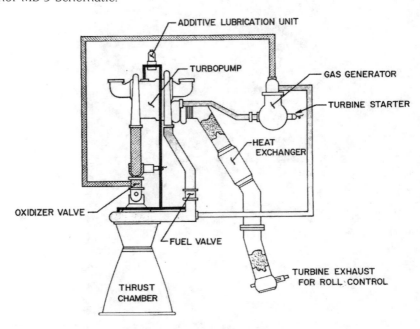

X-1 engine schematic.

The tap-off power cycle was not the only feature demonstrated by the Experimental Engines group that was not adopted for production engines. Anhydrous hydrazine was tested as an alternate fuel for RP-1, but it was more difficult to handle and had no great appeal for the Air Force. More serious consideration was given to spiking up the liquid oxygen with liquid fluorine, a mixture known as FLOX. Up to 20% fluorine was demonstrated to be an easy substitute and was seriously considered by the Air Force as a means to increase the performance of the Atlas. However, when you vented a full tank of FLOX, the fluorine boiled off faster than the oxygen, and so the concentration changed with time. Performance at launch would be a variable, and that was understandably not acceptable to the Air Force ICBM people.

A still wilder idea that came out of Advanced Design was to put the oxidizer pump impeller *inside* the combustion chamber. The reasoning was to eliminate inefficient transfers of energy. In a conventional engine the oxidizer fluid was first given kinetic energy by spinning it up in a centrifugal pump impeller. Then it passed into a diffuser to convert the kinetic energy into potential energy (pressure). At the injector the potential energy was converted back into kinetic energy as the oxidizer went through the injector orifices and then was burned with the fuel. Energy was lost to inefficiencies at each transfer of energy. Why not put the impeller inside the combustion chamber so that the oxidizer gained kinetic energy as it spun up in the impeller and then was burned immediately with fuel with no further conversions of energy? It was interesting reasoning, but this idea was considered just a bit too wild and was never tested.

The Simple H-1 Engine for the Saturn I and IB

Back at Huntsville Wernher von Braun was always looking forward toward his goals of putting humans into orbit and landing them on the Moon and on Mars. For this he was going to need much larger launch vehicles. When Sputnik went into orbit in 1957 with a larger launch vehicle than anything in development in the United States, he saw a sales opportunity. It was still several years before President Kennedy would make his famous speech that started the Apollo program, and so von Braun's proposal would have to be relatively modest to have any chance of getting government funding. He assembled a design team at Huntsville with propulsion support from Rocketdyne. The resulting preliminary design utilized existing propellant tanks with eight Redstone tanks wrapped around a central Jupiter tank core. Propulsion would be from eight engines to be called the 'H-1' that were basically Jupiter engines upgraded in thrust to 165K. The inner four engines would be fixed with the outer four gimbaled to provide attitude control.

Von Braun had finally been sold on the practicality of liquid hydrogen (more on that in Chapter 8), and so an upper stage was envisioned using six Pratt & Whitney RL-10 engines for a total vacuum thrust of 90,000 lb. The name 'Jupiter' had already been used, and so von Braun went to the next outer planet and named the new vehicle 'Saturn I', with the first stage designated as S-I. The second stage would logically have been called S-II, but von Braun had

128   Rocketdyne: Powering Humans into Space

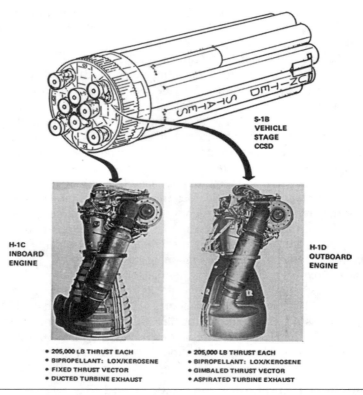

H-1 engine inboard and outboard locations.

designs for vehicles with up to four stages, and so the upper stage for the Saturn I was optimistically called S-IV.

NASA had not yet been formed, and thus the target customer was the Advanced Research Projects Agency (ARPA) in Washington. Von Braun assembled his team for breakfast the morning of the big briefing. He was upbeat, encouraging, and radiating optimism and good humor. He was making jokes, including one where he said, "My proposed big spinning space station, once we get that in orbit, I'm going to invite all the generals in the Pentagon up there. It will be the first time they ever knew which way was up."[7] He thought that was a great play on words and, of course, demonstrated his facility with the American vernacular. In the conference room at ARPA, von Braun gave the introduction, followed by a description of the launch vehicle by his chief preliminary designer, Herman Koelle, then a supporting summary of Rocketdyne's propulsion system by your author. The final pitch by von Braun was very effective, and funding of the project was approved by ARPA, resulting in a contract awarded to Rocketdyne on 11 September 1958 to further the development of the H-1 engine to 188,000 lb of thrust. The eight-engine cluster would now give the Saturn I a takeoff thrust of 1.5 million pounds.

Back at Rocketdyne the Advanced Design people were quick to advocate that all of the simplifying features proved in the X-1 (except tap-off) be adopted for the H-1 engines for the Saturn I. The resulting engine was the most beautifully simple and fail-safe large engine that Rocketdyne had ever put into production. With the turbopump mounted on the side of the thrust chamber, resulting in very short high-pressure propellant lines, the engine was very compact and lightweight, which led to some debates with the engineers at Huntsville. Rocketdyne engineers were not convinced that von Braun's engineers had been equally clever in saving weight. Jack Conyers, senior designer in the Advanced Design group, did a study that demonstrated that at least 2000 lb could be cut from the weight of the engine thrust structure by using a monocoque design rather than Huntsville's rather crude I-beams. Von Braun ended that discussion by stressing that he wanted a thrust mount "so simple that one engineer using a slide rule can analyze all the loads in just five minutes." The result of this design philosophy was a relatively heavy launch vehicle that was no competitor for the later Titan III design, but with this approach von Braun caught up in a hurry with the Soviet launch capability and could start carrying heavy payloads into orbit.

It is appropriate to note again the universal respect that Rocketdyners had for Wernher von Braun, and not just for his ability to promote spaceflight. Theodore "Ted" Benham, who worked with von Braun during engine development for the Redstone, Jupiter, and Saturn launch vehicles, summarized his regard for von Braun as follows, "Excellent to work with. That guy, I had to admire him. We'd get into meetings and he'd call all the guidance people, the ground support people, the test people—he'd get everybody in the same room. And if a guy had a problem, like maybe a guidance fellow had a problem, and the guidance fellow would start telling them what the problem was and what they were doing about it. Well, some of that guidance business is very complicated. But von Braun understood it and he would stop the guy and he would turn it around and put it all into plain English for everybody to understand, and he did this purposely for other people to determine whether any of this affected them or not. And it really worked out to be a very good procedure. . . . He was an awful nice guy, smart as a whip. He worked well with everybody, both the contractors and also the people at Marshall Space Flight Center. He was an exceptional individual."[8]

H-1 Powers Astronauts into Orbit

The first prototype engine was delivered in just seven months, the first launch with only a dummy upper stage was on 27 October 1961, and all 10 flights of the H-1-powered Saturn Is were successful. On one flight an H-1 engine did shut itself off prematurely. However, it was a smooth fail-safe shutdown, and the remaining seven engines easily made up the loss and reached the design cutoff velocity—not a bad record for the H-1s and the Saturn Is.

A seemingly minor but important detail was that all eight engines were not started simultaneously. Analysis had indicated that a simultaneous start would require such an acceleration of propellants that they would cavitate (create bubbles) in the propellant lines. The pumps would then overspeed and fail. The simple cure was to just sequence the starts over a second or two. The same practice was used later for the five F-1 engines on the Saturn V. George Sutton, who has made quite a study of the Russian rocket engines, believes this was the cause of the disastrous launch pad explosions of the gigantic N-1 launch vehicle, their rival to the Saturn V. Sutton believes they went with simultaneous starts and paid the price.

The next step for Saturn I was to upgrade it to a Saturn IB configuration with the H-1 engines up to 205K for a total liftoff thrust of 1.64 million pounds. While the original S-IV upper stage was powered by six Pratt & Whitney RL-10 Centaur engines, the upgraded stage, designated the S-IVB, was powered by a single Rocketdyne J-2 oxygen/hydrogen engine at 230,000 lb of thrust (we will tell the story of the J-2 in Chapter 7). It was a Saturn IB that launched the first manned Apollo spacecraft, Apollo 7, propelling astronauts Wally Schirra, Donn Eisele, and Walter Cunningham into Earth orbit on 11 October 1968 and advancing the United States to neck-and-neck with the Soviet Union in the great space race. Then came the saga of Skylab, a very large workshop in space that was propelled into orbit by a Saturn V on 14 May 1973. This was a major space station built using the structure of a Saturn V upper stage. To appreciate its size, one should see the exhibit at the National Air & Space Museum in Washington. It had the spaciousness of a three-bedroom house and included crew-comfort features like separate private sleeping compartments. Unfortunately a thermal shield broke loose during launch, taking along with it one of the two large solar panels that would power Skylab. The remaining panel was jammed with debris and would not deploy. The human crew to follow had their work cut out for them.

Just 11 days later a Rocketdyne-powered Saturn IB carried Charles "Pete" Conrad, Joe Kerwin, and Paul Weitz to Skylab, where they immediately deployed a hastily designed but ingenious device that enabled them to pass a large sheet of plastic out through a hatch and erect it like an umbrella to shield the Skylab from the intense heat radiating from the sun. Then Conrad and Kerwin, in a four-hour EVA space walk, were able to deploy the jammed solar panel before the Skylab's onboard batteries were fully discharged. This was quite a job by a super crew.

On 28 July a replacement crew of Alan Bean, Owen Garriott, and Jack Lousma were transported by another Saturn IB, followed on 16 November by Gerald Carr, Edward Gibson, and William Pogue. These crews performed serious science observations with the many instruments aboard the Skylab, including a large solar telescope that enabled viewing of solar flares by astronomers Garriott and Gibson. The plan had been to conduct periodic boosts of the Skylab to keep it in orbit for continued experiments, including studying the effects on humans of prolonged stays in the zero-gravity environment of space, but the NASA budget had plummeted after the Apollo program and would no longer permit continued operation of this major space station in orbit. When the third crew via an Apollo reentry capsule landed on 8 February 1974, it marked the end of Skylab, which was allowed to decay out of its orbit and be incinerated in a fiery reentry into a remote area of Australia. Space station science would have to be left to the Russian Salyuts and Mir until the beginning of the 21st century and the assembly of the International Space Station.

The ever reliable Saturn IB was called on one last time on 15 July 1975 to carry astronauts Tom Stafford, Vance Brand, and Deke Slayton to their Apollo Soyuz handshake in orbit with Cosmonauts Leonov and Kubasov. For all 10 launches of the Saturn I and nine launches of the Saturn IB, including the carrying of 15 astronauts into orbit, the 152 Rocketdyne H-1 engines performed flawlessly—a triumph in the continuing Rocketdyne campaign for engine simplicity and reliability.

Pictorial engine comparison: A-4, A-6, 120K Navaho, and H-1.

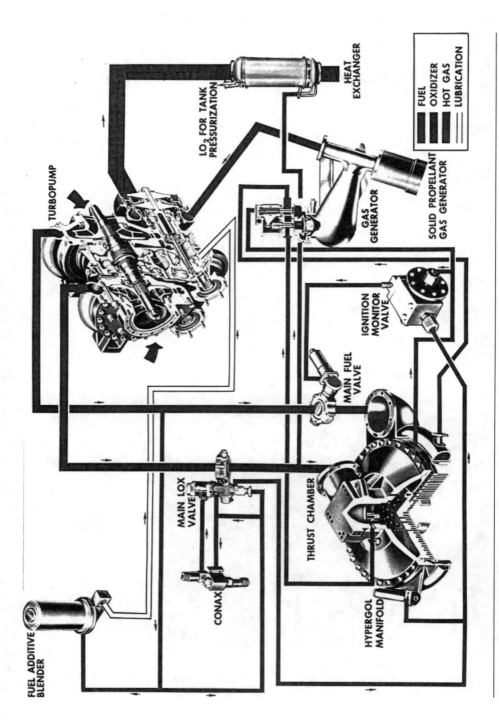

H-1 engine schematic.

Rocketdyne Neosho, Missouri, H-1 engine production line.

Rocketdyne, Neosho, large engine test stand.

S-IB vehicle assembly tank installation at Michoud Assembly Facility (MAF), New Orleans, Louisiana, 1967. The S-1 and S-1B launch vehicle first stages were built by the Chrysler Corporation at MAF, tested at NASA MSFC, Huntsville, Alabama, and Barged to Cape Canaveral for launch.

S-IB vehicle assembly engine closeout at MAF, 1967.

Saturn S-IV stage (six RL-10 engines), 1964.
S-IV and S-IVB stages were assembled at the Douglas Space Systems Center (later McDonnell Douglas) at Huntington Beach, California. Vehicles are approximately 58 feet long and 22 feet in diameter. Hot-fire testing was done at the Douglas Sacramento, California, facility with subsequent shipment to Cape Canaveral or Kennedy Space Center, depending on Saturn I, IB, or V application.

Saturn S-IVB (One J-2 Engine).

An interesting comment comes from Vince Wheelock, who spent 38 years at Rocketdyne in field engineering at customer sites. He says that the compact packaging of the H-1 engines made leak and functional checkouts difficult. The high-pressure lines were too rigid for easy hardware replacement or system isolation. In all other categories the H-1s got top grades.

Because a total of 294 H-1 engines were delivered, there were many spare engines left in storage after the Saturn launches. Happily, these were not wasted, but were adapted to fit the Thor/Delta launch vehicles. For the Saturn application the H-1 engines utilized the compact packaging of mounting the turbopump directly on the side of the thrust chamber. This resulted in short high-pressure lines from the pumps to the thrust chamber but required for gimbaling that there be flexible propellant inlet lines coming from the launch vehicle tanks. Douglas Aircraft did not want to make this change to their Thor/Deltas, and so the surplus H-1s had to be repackaged. Redesignated as RS-27 engines, the repackaged H-1s continued their unblemished record of reliability.

Saturn IB launch.

Skylab Saturn V and Saturn IB on launch pads, 1973.

Nuclear Rocket Paves Path to Hydrogen/Oxygen J-2 Engine

As noted before, a rocket engine is a reaction device, with thrust produced as a reaction to the acceleration of hot exhaust gases. The greater the exhaust velocity, the greater the change in momentum, and the greater the reaction thrust and efficiency, or specific impulse, expressed as "seconds" (pounds of thrust per lb/s of propellant expended). The exhaust velocity can be readily derived to be proportional to the square root of the initial gas temperature divided by the gas molecular weight, $(T/M)^{1/2}$. Therefore, there are two obvious routes to higher specific impulse: 1) increase the initial gas temperature in the combustion chamber, and 2) reduce the molecular weight of the exhaust gas. With chemical combustion rockets one of the very best propellant combinations is hydrogen/oxygen. The combustion temperature of fully reacted O_2 and H_2 is high, and the exhaust product of water has a relatively low molecular weight of just 18. Compare that to a fuel high in carbon where the exhaust gas will be heavy with CO_2 and its molecular weight of 44. Thus you want as much of that light hydrogen in the exhaust gas as possible.

That raised the question by mid-twentieth century, "Why not all hydrogen?" Once nuclear fission had been demonstrated in the 1940s, the concept was developed to use a solid-core nuclear reactor to heat hydrogen and then expel it through a rocket nozzle. While not reaching the combustion level of 6000°F of top combustion rockets, by employing graphite for the reactor core material the gas temperature exiting the core should reach at least 5000°R (more than 4500°F). With the H_2 exhaust gas having a molecular weight of only 2, the nuclear thermal rocket engine should achieve twice the specific impulse of a hydrogen/oxygen combustion rocket.

Development Begins on a Nuclear Rocket

In 1955 both the Atomic Energy Commission (AEC) and the USAF developed interest in this concept as a means to achieve an intercontinental ballistic missile using just a single stage. The AEC started reactor studies at both their Los Alamos Scientific Laboratory (LASL) and Lawrence Livermore Laboratories

(LLL). LASL won that competition and started building a series of graphite-core reactors under the code name KIWI. At the same time the USAF funded studies of liquid hydrogen pumping at Rocketdyne, where engineers proposed a turbine-driven multistage axial flow pump. The Air Force liked the design and gave Rocketdyne a contract for its development. The resulting pump, designated the MK9, could readily pump 10,000 gal/min of liquid hydrogen to 1500 psi. Rocketdyne was also contracted to build high-expansion-ratio nozzles, initially water cooled and then regeneratively cooled with the liquid hydrogen fuel. All of this effort at Rocketdyne in support of a nuclear engine was under the program direction of Stanley Gunn, who had majored in rocket and jet propulsion with a minor in nuclear physics at Purdue University under the renowned Maurice Zucrow. Gunn was not only technically strong but also was an enthusiastic team leader, and the development effort progressed rapidly. Stan was later joined in the nuclear office by Robert Dillaway.

Working with liquid hydrogen was a new experience. Its very low density and low viscosity make it very leaky stuff. Where possible, it is better to weld joints rather than depend on seals. Some metals were embrittled by the hydrogen, but certain stainless steels were found to be fine. Practical lessons had to be learned about working with something whose vapor was lighter than air. During one pump test a nearby tool cabinet in the test cell blew up. Hydrogen vapor had accumulated inside the top of the cabinet and was ignited by a spark—hydrogen is explosive over a wide range of mixtures with air, and it demolished the cabinet. Holes had to be drilled in all of the cabinets and roof openings provided to vent everything *upward*. You had to think about draining up, not down. During gas generator tests with hydrogen and oxygen, special care had to be taken to cordon off the area. The hot exhaust was absolutely clear, with the only visible evidence of its presence being a shimmering of the air. One could easily walk right into it. The test crew finally started adding a touch of methane just to make the exhaust luminous.

The Rocketdyne H_2 cooled nozzle and a gas generator powered MK9 turbopump (the turbine was adapted from the Atlas booster turbopump) were then assembled by Aerojet with the KIWI-B 1000-MW reactor at the AEC's Nevada test site and the entire breadboard engine fired with complete success for 8 min at a chamber pressure of 500 psia and an impressive 50K thrust level.[1] It was a clear demonstration of not only the feasibility of nuclear rocket engines but also the practicality of liquid hydrogen as a rocket fuel.

When NASA was established in 1958, the nuclear rocket program direction was placed under a new Space Nuclear Propulsion Office (SNPO) jointly staffed and funded by NASA and the AEC.[2] The potential application was now focused on sending humans to Mars rather than an ICBM for the Air Force. Harry Finger was the head of the NASA contingent and led a strong effort toward the development of a 75K flight engine called Nuclear Engine for Rocket Vehicle Application (NERVA), but NASA eventually in 1972 was forced to drop the program from its tightening budget. There were immediate and urgent NASA mission needs while Congressional approval of a manned mission to Mars was clearly many years in the future. The entire nuclear engine program had been an impressive technical success, but it was before its time. While the application of nuclear engines may have been delayed, liquid hydrogen was there to stay.

Selling Hydrogen/Oxygen

In 1957 the Advanced Design group at Rocketdyne, with encouragement from John Tormey's research group, put together a comprehensive briefing that summarized the impressive progress in engine component development for liquid hydrogen and demonstrated its great performance advantages as a fuel for future missiles and launch vehicles, especially for upper stages. The story was presented to Tom Dixon, Rocketdyne's Vice President for Development. Tom, always receptive to new advances, was more than just enthusiastic. He went on a crusade to spread this new revelation to the Air Force.

With your author in tow, he headed for the Pentagon. Zooming from office to office, he would ask each official how much time they could spare to hear a story vital to the nation's future. The full story took about an hour, but given 20 minutes, he would say, "Okay, Bob, 20-minute summary." In some offices he was given only 10 minutes, or even just five, but he was never turned away. Tom was even able to walk into the office of the Secretary of Defense without an appointment, and when the Secretary heard it was Tom Dixon with an important message, he knew Tom well enough that it must be something Tom sincerely believed to be vital, and he interrupted his packed schedule to make 10 minutes available. That Tom Dixon was so respected that he could get that kind of access was really amazing.

Returning to Canoga Park, Tom did not wait for substantial amounts of money to arrive from the Air Force. He gathered a team of doers from Advanced Design and the engineers working with liquid hydrogen, like Stan Gunn, plus Paul Castenholz, the development engineer who was well known for getting experimental engines built and tested in a hurry. Tom launched into a pep talk on the need to start design and development on a new hydrogen/oxygen rocket engine, immediately if not sooner. Out of the air he picked a thrust level of 200K—no particular application, it just sounded like the right level to him. He would prove to be a good prophet.

Tom always put the emphasis on *"you."* He never said, *"I want this done."* It was always, *"You can do it."* When asked about funding, he ignored the question and repeated that the country needed this engine and *"You can do it."* We assumed he meant to use company funds. Castenholz reminded him that the engine test stands belonged to the government, and that the company would be liable for any damage if we blew up an engine. Again the question was ignored. *"You can do it."* Tom may not have been the world's most practical business manager, but he was inspirational. Thus the team charged off and got to work. Stan Gunn, who had directed the hydrogen turbopump and nozzle development for the nuclear rocket, was appointed the Program Manager and work got rapidly under way. The ever-responsive Air Force propulsion people soon picked up the funding at a modest level. The MK9

MK9 liquid hydrogen turbopump.

Wernher von Braun SSFL with Rocketdyne President, Sam Hoffman, at Santa Susana being briefed by Stan Gunn on Rocketdyne's liquid hydrogen research, 1957.

axial hydrogen pump was found to be very suitable, and so the new 200K engine had a great head start. Tests of a new 200K thrust chamber were soon under way.

About that time Rocketdyne had a visit from Krafft Ehricke, a rocket engineer well known for his exceptional analytical ability and boundless energy. Krafft had been a junior member of von Braun's Peenemunde crew and after the war had joined Gen. Walter Dornberger at Bell Aircraft. In 1954 he moved to Convair in San Diego and soon began developing a design for a high-performance upper stage he called 'Centaur.' At Rocketdyne he met with the Advanced Design people and explored the feasibility of using liquid hydrogen with liquid oxygen as propellants. He got an enthusiastic response and was told of work well in progress with liquid hydrogen. He returned in a few months with a hydrogen/oxygen stage design. Strangely enough, it employed pressurized propellant tanks rather than a turbopump. With the low density of liquid hydrogen, a pressurized tank is quite heavy, but Krafft thought it would be easier to sell if he kept the design simple. Rocketdyne thus provided him with the preliminary design of a pressurized hydrogen/oxygen propulsion system.

Krafft carried his design and development pitch to Washington to present to the Department of Defense's Advanced Research Project Agency (ARPA). The propulsion part of the presentation was given by your author. Krafft's story was well presented, and his enthusiasm was infectious. The response of ARPA head Roy Johnson and his crew was very positive. Back at Rocketdyne the Preliminary Design group began preparation of their proposal for the Centaur's engine. With all of the hydrogen/oxygen work going on for NERVA and the 200K engine, there was just no way Rocketdyne would not win over Aerojet in the coming competition.

Then in October 1958 Col. Norman Appold, detailed from the USAF to ARPA, announced that he was awarding the hydrogen/oxygen development

Krafft Ehricke

contract *sole-source* to Pratt & Whitney! No competition? Pratt & Whitney had never built any kind of rocket engine, small or large. They were a turbojet engine company. There was total disbelief at Rocketdyne, quickly followed by outrage. A formal protest was seriously considered, but in those days you just did not protest against your very best customer. Appold's excuse came out only later. Pratt & Whitney had been developing a hydrogen pump for a SECRET hydrogen-fueled jet engine. However, Rocketdyne was well along in the development of a hydrogen pump and hydrogen-cooled nozzles for the nuclear rocket and had fired hydrogen/oxygen rocket motors, plus had more than a decade of experience in large rocket engines. Rocketdyne old-timers, even today, get agitated and flushed when they start reminiscing about the injustice of that award. Years later Dick Mulready of Pratt & Whitney would write a book in which he expressed his shock that Rocketdyne could have overcome the lead that Pratt & Whitney held in staged combustion and high chamber pressure to win the Space Shuttle main engine (SSME) competition.[3] Well, if anything, the shock at Rocketdyne in October 1958 was even greater.

The J-2 Is Born

The sense of outrage just accelerated the work at Rocketdyne on the new 200K hydrogen/oxygen engine, which was named the J-2 (after the H-1, Rocketdyne skipped over 'I' as the capital "eye" looked too much like the numeral 'one'). Following their continuing aim of making engines simple, the Rocketdyne engineers continued with a gas generator cycle, making engine design and development very straightforward. Development was progressing at a moderate pace under Air Force funding and then made a smooth transition to a NASA contract in June 1960. MSFC was appropriately assigned NASA management of the contract as the J-2 was going to fit right into von Braun's latest plans for sending humans to the Moon. The experienced Walt Studhalter was named J-2 Program Manager at Rocketdyne, to be later succeeded by Norm Reuel with Paul Castenholz as his deputy. Willie Wilhelm, who had been chief project engineer for the very successful Thor engine, was switched over to be chief project engineer on the J-2.[4] Norm Reuel had responsibility for development of all of Rocketdyne's large engines, and he was becoming limited by heart problems, and so it was Castenholz who was really directing day-to-day J-2 engine development. He was soon promoted to be the J-2 program manager with Willie Wilhelm and later Paul Fuller as his deputy.

It had taken a long time to convince von Braun of the practicality of employing liquid hydrogen. Rocketdyners had always found him to be "positive, very adventuresome, and interested in new development,"[5] but he had heard horror tales of the difficulty of sealing the very low density and low viscosity liquid hydrogen, as well as the phenomenon of hydrogen embrittlement of common metals and alloys, and so he had serious doubts about its practicality. When NASA had picked up responsibility for the Centaur stage from ARPA, they assigned project management initially to the Marshall Space Flight Center, but there it was given very low priority, and in 1962 the project was

Walt Studhalter   Norm Reuel   Paul Castenholz

Willie Wilhelm   Paul Fuller

transferred to the Lewis Research Center in Cleveland, where they had experience with liquid hydrogen and were enthusiastic about its application.

Several groups and individuals claim credit for swinging von Braun over to hydrogen. John Sloop and his people at Lewis said they briefed him on their good experience with liquid hydrogen, but it is not likely that Wernher was too receptive at that time. A more important event was probably a visit that he made to Rocketdyne where Doug Hege had arranged with Stan Gunn to give a presentation on the work under way for NERVA and the 200K engine.[6] He was shown the welded joints that eliminated leakage, was presented a list of stainless steel alloys that had demonstrated no hydrogen embrittlement, was updated on the good progress toward the 200K engine, and most importantly was given a demonstration of the successful MK9 turbopump. Von Braun said he was very impressed and thereafter became much more receptive to the use of liquid hydrogen. The final factor in converting von Braun was undoubtedly the forceful influence of Abe Silverstein at NASA Headquarters. As John Sloop's former boss, Abe had long been sold on the merits of liquid hydrogen as a rocket fuel. Later Pratt & Whitney gave von Braun a hot-firing demonstration of their RL10 Centaur engine, but he was already sold by then. In December 1959 NASA Headquarters made the final decision to use hydrogen/oxygen in the upper stages of the Saturn launch vehicles.

Development of the J-2 engine went very smoothly. The hydrogen turbopump known as the MK15 was a direct adaptation of the MK9 unit already developed for the nuclear rocket program. Reflecting the low density of liquid hydrogen, it was a seven-stage axial pump driven at a very high speed of 27,270 rpm by a two-stage velocity-compounded turbine on the same shaft. The liquid oxygen (LOX) pump was a more conventional single-stage centrifugal design driven at 8690 rpm by its own two-stage velocity-compounded turbine. Both pumps had inducers to eliminate any cavitation. Neither pump required any external fluid for bearing lubrication or cooling. The pumps were compactly mounted on opposite sides of the thrust chamber with flexible axial inlets. A single bipropellant gas generator provided fuel-rich gas to drive the turbines.

Rocketdyne stuck to its principles in keeping the engine as simple as possible. However, the requirement for restarts in space necessarily added some complexity. A 4.2-ft^3 spherical start tank was added to supply high-pressure gaseous hydrogen to spin up the turbines for engine start. As the hydrogen and oxygen pump discharge pressures built up, the system bootstrapped itself to its full thrust level. The start tank was refilled in flight for subsequent engine restarts. Electric spark igniters were provided in the bipropellant gas generator and the main chamber. High-pressure gas from the start tank and then from the gas generator was ducted first through the fuel pump turbine, then to the oxidizer pump turbine, and then to a heat exchanger that could heat either oxygen or helium to pressurize the vehicle's oxidizer tank. The turbine exhaust gas was then injected into the thrust chamber nozzle at the 16:1 expansion point where it combined with the main exhaust gas and contributed to the net thrust. Heated hydrogen could be tapped off the thrust chamber cooling jacket to pressurize the vehicle fuel tank. Helium provided a convenient fluid for actuating the very cold main valves. A bypass valve on the LOX pump provided a moderate degree of mixture ratio and thrust control. The starting procedure had to be carefully developed—too rapid a start tended to produce cavitation and stalling in the axial-flow hydrogen pump. Engineer Keith Watts is credited with working out the successful sequencing of the propellant valves. The resulting system was not as simple as the "ultimate" H-1, but still quite straightforward for a high-performance restartable engine.

Thrust chamber development progressed rapidly—hydrogen was already known to be an excellent coolant. Using tapered tubes allowed the ready fabrication of the 27.5:1 high-expansion-ratio nozzle. In the vacuum of space the resulting net thrust was 230,000 lb. It is interesting to compare the J-2 performance to that of the Atlas 60K sustainer engine. Both had essentially the same chamber pressure (717 vs 706 psia) and nearly the same nozzle area ratio (27.5:1 vs 25:1) so that the huge increase in vacuum specific impulse (425 vs 307 s) is due to advancing to liquid hydrogen as the fuel instead of RP-1 kerosene.

The 27.5:1 nozzle presented an interesting problem during ground tests. At sea-level pressure the nozzle flow separates from the nozzle walls before the nozzle exit. The separation is not always perfectly symmetrical, producing substantial side loads on the nozzle. This is not a problem for the full-flowing nozzle during its in-flight altitude operation. To handle the side loads, a special nozzle support strut was adapted for use during the ground-firing tests. To verify altitude performance, a J-2 engine was shipped to the Arnold Engineering Development Center (AEDC) and fired in their large vacuum facility.

Happily, there were never combustion instability problems with hydrogen/oxygen. The heated hydrogen reaches the injector in a gaseous state so that a coaxial injection pattern was used with the hydrogen gas surrounding each oxygen jet. The combustion was so rapid that there was no chance for appreciable amounts of unburned propellants to accumulate in the combustion chamber and add energy to any acoustic oscillations. Hot-firing tests presented a new experience to observers in that the exhaust was translucent rather than the bright orange flame produced by hydrocarbon fuels. There were brighter "Mach diamonds" visible in the exhaust marking the location of the expansion and compression waves produced by overexpansion and separation in the high-expansion-ratio nozzle. A night firing was a thing of beauty, with the translucent exhaust plume now visible against a dark background.

Engine development passed all of its milestones with relative ease. The Rocketdyne team did not need any extra motivation, but they were proud of the fact that they earned maximum award fees under their NASA incentive contract. To formally qualify for manned flight, the J-2 was required by NASA to complete 69 consecutive successes. It did that in March 1966. Although not required to, Rocketdyne continued that test series for 220 consecutive successes, a demonstrated engine reliability of 99.8%.[7]

Reliable J-2s Power S-IVBs into Orbit

The von Braun team at MSFC had designed their Saturn I launch vehicle using eight of Rocketdyne's oxygen/RP-1 H-1 engines for the booster stage. The upper stages were to utilize hydrogen and oxygen. Initially the second stage built by Douglas Aircraft and called S-IV was powered by six Pratt & Whitney RL10 engines for a total vacuum thrust of 90,000 lb. That configuration made six successful developmental flights. The second stage design was then advanced to the S-IVB version, which utilized a single J-2 engine with its vacuum thrust of 230,000 lb. This launch vehicle configuration was identified as the Saturn IB. The J-2s operated flawlessly for all nine launches of the Saturn IB, performing developmental flights (both manned and unmanned) for the Apollo program, delivering three 3-man crews to the Skylab space station, and culminating in the milestone Apollo Soyuz mission with its handshake in orbit in 1975.

J-2s Power Upper Stages for Saturn V

Beyond the Saturn I and IB, MSFC was working on an even larger launch vehicle that would become the giant Saturn V with its boost stage powered by five of Rocketdyne's F-1 engines for a takeoff thrust of 7.5 million pounds (see Chapter 8 for the history of the F-1). Even before the Saturn V design was firmed up, von Braun issued a Request for Proposal on its second stage, designated the S-II. The fierce competition was won by North American Aviation with a design that featured an aluminum alloy that increased in strength at cryogenic temperatures, resulting in a very lightweight stage.[8] That winning S-II design utilized five J-2 engines for a total vacuum thrust of 1.15 million pounds.

In his historic speech of 25 May 1961, President Kennedy committed the United States to land a man on the Moon before 1970. In the fall of 1963, George Mueller had succeeded Dale Myers as NASA's head of manned space flight, and he calculated that NASA would not make that ambitious schedule unless it were willing to gamble on its schedule of flight testing. Von Braun had planned the normal series of test flights for the Saturn V, just as he had for the Saturn I, starting with a launch of the booster stage only with dummy upper stages. That would be followed by a launch that added the second stage. Then there would be a launch with all-up three stages, the S-IC, the S-II, and the S-IVB. With this approach any flaws in the stages could be corrected before risking adding a new and untried stage. However, Mueller had successfully gambled on an all-up test of the Minuteman ICBM when he was leading that project, and he directed doing the same with the Saturn V. Von Braun pointed out that they would look like reckless gamblers if anything failed on that all-up launch. Mueller would no doubt have been replaced in his job for taking such an expensive gamble, but George must have been living right and on good terms with his guardian angel because the first flight of the all-up three-stage Saturn V on 9 November 1967 was a glorious success, and Apollo was on schedule for a landing on the moon in 1969. In all fairness, the full credit should not go to George or his guardian angel. Let us acknowledge the superb and dedicated work of many hundreds of engineers, machinists, and technicians at Boeing, Douglas, North American Rockwell (including Rocketdyne), IBM, and all of their subcontractors.

Solving a Surprise J-2 Shutdown

After that great beginning, the next Saturn V launch on 4 April 1968 produced a shock. One of the J-2 engines in the S-II stage began to experience a decay in its chamber pressure and then lost all pressure and shut down after 263 seconds. Then the single J-2 in the S-IVB stage also experienced a decay in pressure, although it continued to fire for its planned duration and enough more to make it into orbit—but then it refused to respond to commands for a planned restart. Although orbit was achieved, no one was going to have astronauts trust their lives on the erratic performance of the J-2s. What had happened to that demonstrated 99.8% reliability?

Astronauts were planned to be aboard a Saturn V on its very next launch, and so all of the managers of the entire Apollo program had come to the Kennedy Space Center to witness the critical second launch. Paul Castenholz remembers well the post-launch meeting, "At the end of the flight we went into a big meeting room at Cape Canaveral. I said 'I have to tell you we don't know what happened. But we will immediately go on a 24-hour schedule, and we will keep you informed.'" Paul recalls, "It was probably my lowest point as a rocket engineer. I was thinking, 'What could be worse than this?'"[9]

Castenholz has been described variously as athletic, casual, and calm but also as aggressive, intense, and hard driving. What really counted and made him a super development engineer was that he took quick and determined action to solve problems. If paperwork or cumbersome procedures got in his way, he either bulldozed through them or else just went around them. Here he and his J-2 team were facing a critical problem with scant evidence to go on. There had never been a thrust decay problem during hundreds of static firings, and J-2s had performed well on all earlier flights in the Saturn IB series as well as the prior Saturn V flight. Was there some new element in this latest flight? The J-2 was considered to be fully developed, and so there had been little telemetry of engine parameters. The S-II stage had burned up on reentry while the S-IVB stage was up in orbit. The net result was that there was no hardware to examine and little information to go on. However, detailed examination of the limited telemetry did indicate that the culprit was a ruptured fuel line, very likely the small flexible line that carried fuel to the engine igniter. A rupture of that line would explain the engine's failure to restart.

Firing tests were immediately conducted putting the suspect line under increased stress. It held up well under all conditions—stalemate. A key turning point was when team member Marshall McClure raised the question, "Would it be different in space than on the ground?"[10] A restudy of films of J-2 tests showed that considerable ice was building up on the cryogenic oxygen and hydrogen lines. In space there would be no water vapor and therefore no in-flight ice buildup. The suspect fuel line incorporated a flexible bellows. During engine operation on the ground, ice formed quickly in the corrugations of the bellows and was likely very effective in damping out any vibrations. With around-the-clock shifts, a special vacuum tank was built in just two weeks. Eight fuel lines were placed in the vacuum tank where no ice could form and were subjected to flight loads and vibrations. All eight ruptured. Conclusion: the bellows were marginal in strength and held on some flights but were subject to failure. The solution was to switch to a solid line, with bends to provide for the flexing, and the problem never reappeared again.

The next Saturn V launch on 21 December 1968, with the mission designation Apollo 8 and with astronauts Frank Borman, James Lovell, and William Anders aboard, went off without a hitch. The detective work and reasoning had been brilliant, and Castenholz was now held in high esteem at MSFC, which would have impact later in the all-important SSME competition (see Chapter 9).

There was some pressure oscillation experienced in the second stage on some flights. A low frequency oscillation, typically 5–20 Hz, was experienced

Paul Castenholz, standing, J-2 Program Manager, holds new Augmented Spark Ignition (ASI) line for J-2 engine while Paul Fuller, J-2 Project Engineer, Inspects its location. More than 1000 Engineers and scientists teamed up to discover the cause of the malfunction in the ASI fuel on the Apollo 6 flight.

occasionally in the propellant system of most large launch vehicles. Because it was likened to the motion of a child bouncing on a pogo stick, the phenomenon was known as 'pogo'. While not usually considered critical, it was of concern. During the launch of Apollo 13, pogo occurred during the second-stage burn, of sufficient duration to trigger a chamber pressure sensor that shut down the center J-2 engine. The outer four engines readily made up for the lost impulse, and there was no impact on the mission. The problem was in the plumbing of the S-II stage, developed by Rockwell's Space Division, and their engineers solved the problem for subsequent flights by the insertion of a helium-bleed pogo suppressor in the oxidizer plumbing.[11]

Rocketdyne delivered 152 of the J-2 engines to NASA. Eighty-six were flown, and except for that one test flight and the one premature shutdown due to pogo in the second-stage plumbing, all performed superbly, including launching astronauts to the Skylab space station, carrying Stafford, Brand, and Slayton to their historic docking with Soyuz, and sending all of the Apollo astronauts to the Moon. It was indeed an engine to be proud of.

Nuclear Rocket Paves Path to Hydrogen/Oxygen J-2 Engine

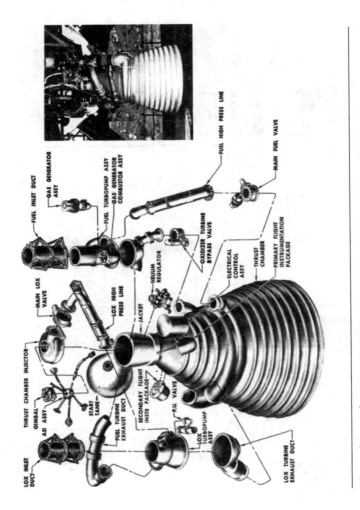

J-2 engine major components.

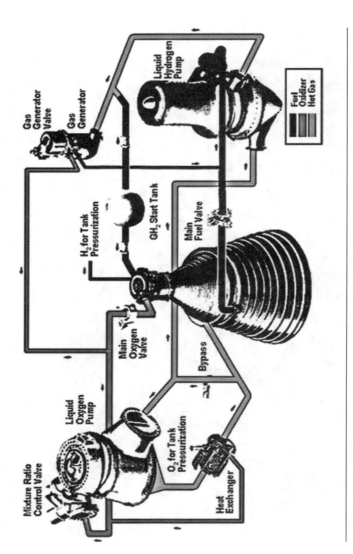

J-2 engine schematic.

Nuclear Rocket Paves Path to Hydrogen/Oxygen J-2 Engine 153

J-2 engine thrust chamber assembly initial cooling tube placement in form of thrust chamber.

J-2 engine thrust chamber final cooling tube placement.

J-2 engine thrust chamber assemblies are completed in "white room" at Canoga Park. The assemblies then go to gas-fired furnaces where the tubes are brazed together.

J-2 engine thrust chamber assembly closeout.

J-2 engine main injector being displayed by Paul Castenholz, J-2 engine program manager. The injector contains 640 orifice sets, and the thrust chamber behind Paul has 540 cooling tubes.

J-2 engine fuel turbopump rotor assembly.

J-2 engine final assembly line. The J-2 engine has approximately 1700 different parts with total multiple usage of more than 10,000.

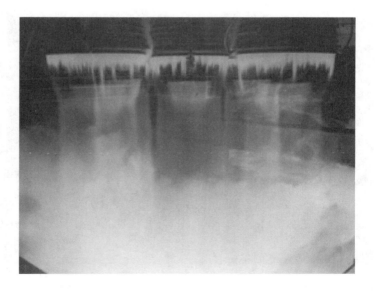

A five-engine J-2 cluster hot-fire picture taken at the Rocketdyne Santa Susana Test Facility during a test of the S-II battleship Stage.

J-2 engine, in a rarely seen view, looking up thrust chamber exit, when installed in test stand at the Santa Susana facility.

The Arnold Engineering Development Center (AEDC) Rocket engine development test cell J-4 is used to test many large rocket engines including, in the past, the Rocketdyne Apollo Saturn J-2 engine. Altitudes up to 100,000 FT are simulated. Pictured is an artist's concept of the cell.

Saturn V second stage (S-II) aft. The S-II stage was built by North American Rockwell's Space Division (initially North American's Space & Information Systems Division). Design was accomplished in the Downey, California, plant with assembly at the Seal Beach, California, plant. It was the largest hydrogen-fueled stage ever built being 81.5 FT tall, 33 FT in diameter, and weighed more than one million pounds when fueled. Hot-fire testing was done at the Mississippi Test Facility (MTF) with subsequent barging to Kennedy Space Center (KSC), Florida, for launch. Pictured is the S-II stage with the five J-2s.

Performance & Weight

Nominal vacuum thrust (lb)	230,000
Nominal vacuum specific impulse	426
Chamber pressure (psia) (nozzle stagnation)	717
Engine mixture ratio calibration (O/F)	5.5:1
Basic engine dry weight (lb)	2,754
Engine dry weight (lb) (including accessories)	3,492

Description

Pump-fed, liquid-propellant rocket engine
Propellants: liquid oxygen & liquid hydrogen
Nozzle area ratio: 27.5:1
Tubular-wall thrust chamber, regeneratively cooled
Separate oxidizer & fuel turbopumps
Bearing lubrication: liquid oxygen & liquid hydrogen
Turbine drive: gas generator burning main propellants

34 Flights, 86 Engines, 98% Reliability

J-2 Basic engine characteristics.

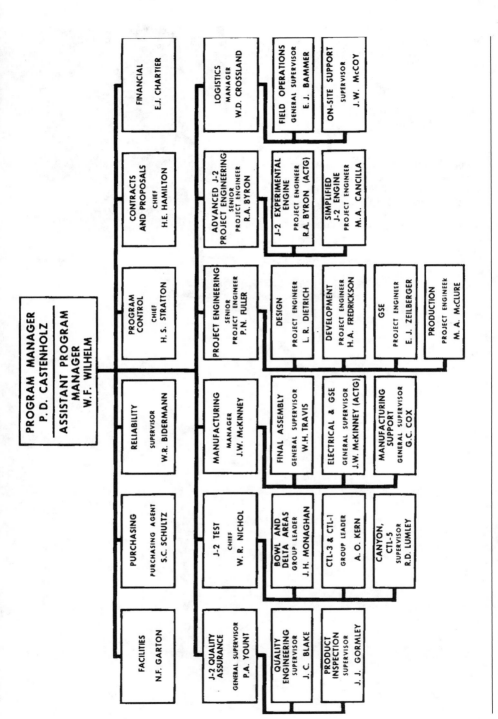

Rocketdyne J-2 program organization.

CHAPTER 8

F-1s and J-2s Power Astronauts to the Moon

In 1955 the engineers at the USAF Propulsion Laboratory raised a good question, "How big can you make a liquid-propellant rocket engine—what is the limit to the thrust level of an individual engine?" To answer that fascinating question, the Advanced Design group decided to bypass the tedious process of looking for limits. That approach would involve months of detailed analysis into each and every engine component looking for fundamental limiting factors in all phases of engine operation and fabrication. Instead it was decided to pick the highest single-engine thrust level anyone could imagine, do a preliminary design of that particular engine, and look for limits in that specific design.

The Bold Advance to 1,000,000 lb Thrust

Detailed design work was already under way on a 300–400,000-lb thrust engine designated the E-1. Herman Koelle, head of preliminary design for von Braun's team, had already declared the E-1 to be the largest engine that would ever need to be built, but design work had found no limits to that size engine. Putting heads together, the preliminary design team asked everyone to name the very highest thrust level they could possibly imagine at any time in the future—forget any application, just imagine a level sometime in the far future. The very highest number anyone could come up with was 1,000,000 lb. Anything beyond that seemed like just science fiction. A preliminary design was done for such an engine, with particular attention to the turbopump, where it was conjectured that stress limits would be encountered. No such limits were found. When the results were presented to the Air Force in 1957, they said, "Great. Go ahead and build one." Initial funding was modest but covered detail design and the fabrication and testing of an uncooled thrust chamber at the million-pound thrust level. Ah, those were days of bold decisions and actions.

Moving up the alphabet, the new engine was named the F-1. Program management was put under Dave Aldrich, who was managing new technology development under the REAP program. Dom Sanchini was selected as his deputy, and Bob Linse was the senior project engineer. As the project progressed, Ted Benham was added as the F-1 development project engineer and would later succeed Linse as the senior project engineer, with Stuart "Stu" Mulliken stepping up into the role of development project engineer.[1]

While detailed design was progressing on the turbopump, an uncooled "boiler plate" thrust chamber was built to start testing. Because it was not cooled, it was limited to short-duration start tests. The chamber was a monster in size compared to anything seen before and was quickly dubbed the "King Kong." When it was fired at Santa Susana, it lit up the sky and rattled all of the dishes in the San Fernando Valley, especially when there was an atmospheric inversion layer that reflected the sound waves back down to the surface. The Public Relations office had to step up their community relations efforts, calling it the "sound of freedom," with the implication that this was part of a vital national defense project. There were really no serious protests.

While a stable thrust level of 1,000,000 lb was indeed realized in 1959, the chamber was very prone to rough starts and destructive combustion instability. It was clear that full development was going to be a major challenge. With no application in sight, there was little motivation for the Air Force to continue at any rapid pace. Nevertheless, on 23 June 1958 the USAF awarded a contract to Rocketdyne to proceed with the design of a full F-1 engine capable of producing up to 1.5 million pounds of thrust. (How were the USAF people able to foresee that NASA would soon want 1.5 million pounds of thrust rather than the already bold 1.0 million pounds? The record is not clear.) ARPA got into the act via their charter to promote needed new technology for all of the armed services, and they apparently were the driving force to reassign F-1 project management responsibility to von Braun's ABMA. This was fine with the Rocketdyners, who had excellent working relations with the Huntsville team. Then NASA became an official entity on 1 October 1958, and based on extensive studies of manned missions to Earth orbit and to the Moon conducted by a team at the Langley Research Center, they knew that they would need a very large launch vehicle.

Very quickly on 21 October 1958, just three weeks after they became official, NASA issued a request for proposal to Rocketdyne for the development of a million-pound thrust engine. Rocketdyne responded promptly on 10 November 1958 with a proposal signed by Tom Dixon. For a formal government procurement, things were really moving quickly. The proposal was accepted, and NASA signed a contract for development of the F-1 engine on 19 January 1959.[2]

The F-1 Raised to 1,500,000 lb Thrust

Abraham "Abe" Silverstein was responsible at NASA Headquarters for all development, and he insisted that the thrust level be increased to 1.5 million pounds to equal the original design takeoff thrust of the Saturn I. Abe came from the NACA Lewis Research Center in Cleveland and was highly respected for his technical direction in the development of both aircraft engines and rocket engines. John Sloop, a pioneer in the development of hydrogen/oxygen rocket engines and an associate of the colorful Silverstein, described him as "sharp, imaginative, aggressive and decisive. He was a hard bargainer at the conference table but very warm-hearted in personal relationships."[3] Abe was to play a major role in the early days of NASA.

On test facilities things moved equally fast. King Kong had rattled so many dishes in the San Fernando Valley that it was clear that F-1 testing would have to be more remote. Construction began in February 1959 on three test stands out in the Mojave Desert on a ridge that was in a remote area of Edwards Air Force Base. Three more stands were added later. The complex would become known as the Edwards Rocket Base (ERB).

As the detailed design of the F-1 engine was firming up, Abe Silverstein was free with his advice. Without argument his years of experience with both aircraft engines and rocket engines entitled him to some strong opinions. He knew that going to 1,500,000 lb of thrust was going to be a challenge. He specified that the F-1 be based on conservative existing state of the art and that it must demonstrate the high reliability required for human flight. This fit well with the thinking at Rocketdyne. From their very beginning in 1945, the Rocketdyne engineers had pursued two main goals: 1) to steadily advance the thrust level of their liquid-propellant rocket engines, and 2) to simplify the rather complex subsystems of the pioneering German A-4. From the very beginning the Rocketdyne approach on the F-1 was to follow the lead of the very successful and highly simplified X-1 and H-1 engines.

Designing the F-1

Following a conservative approach, the F-1 thrust chamber was to be of the well-demonstrated regeneratively cooled tubular construction with a bell nozzle. Chamber pressure would be progressed moderately to 982 psia to keep the size down and boost sea-level specific impulse. The turbopump would have both pumps and the turbine mounted on a single shaft, eliminating the need for any gears. Following the H-1 design, the turbopump would be compactly mounted on the thrust chamber. The standard gas generator cycle would be continued with a start achieved with a pyrotechnic charge in the gas generator and just tank head pressure. The turbine exhaust gases would be used to heat helium to pressurize the launch vehicle fuel tank while oxygen would be vaporized to pressurize the oxidizer tank. In the now-standard pressure-ladder start sequence, the RP-1 fuel would be utilized to actuate the main valves, as well as to lubricate the turbopump bearings and to serve as the hydraulic fluid for the gimbal actuators. As fuel pressure built up during start, it would expel a hypergolic mixture (triethylaluminum and triethylboron) into the main chamber to act as the ignition source.

On the design team was a young engineer by the name of Bob Biggs, who had worked on the engines for the Navaho III and the Jupiter missiles. Biggs initiated an analysis of the planned tank head start and pressure ladder sequence for the F-1 engine. With this start sequence the gas generator would

Ted Benham, F-1 Senior Project Engineer, next to early F-1 engine mockup.

Bob Biggs, a young F-1 engine design engineer, shown working modeling techniques for the Jupiter S-3D engine in 1959.

ignite with propellants right from the tanks, and the engine would bootstrap from there up to full thrust. In this sequence, the fuel serves as the hydraulic fluid to actuate the main propellant valves. Biggs was able to show convincingly that with the valves closed the pumps would cavitate as they spun up, and because they were trying to pump vapor instead of liquid, they would overspeed and fail before the main propellant valves could open under the rising fuel discharge pressure. The simple solution was to open the main LOX valves earlier using fluid from a ground source. The resulting liquid oxygen flow through the LOX pump impeller started the turbopump spinning and produced a smooth bootstrap to full power. Biggs earned new respect for preventing what would have been a destructive engine failure, and he went on to play a major role in the later development of the SSME (see Chapter 9).[4]

Abe Silverstein added another conservative touch by specifying that there be dual main oxidizer valves and dual main fuel valves. With such a large combustion chamber, he believed that multiple valves might be required to get a proper propellant distribution. He was probably right, as this chamber was far bigger than ever seen before. It was to be calibrated to produce a nominal 1,500,000 lb of thrust, but with turbine exhaust added the F-1 engine's official rated thrust was to be 1,522,000 lb.

The thrust chamber design was quickly firmed up as a fairly standard Rocketdyne design using a bundle of contoured thin-walled Inconel X tubes. One variant was that the bundle started out with 178 tubes at the head end but then was split or bifurcated to 356 tubes about halfway down. The nozzle was a regeneratively cooled bell shape to an expansion ratio of 10:1. A single-wall nozzle extension cooled by injecting the turbine exhaust gas carried the expansion to 16:1. This extension could be removed to facilitate transportation of the engine. The world's largest gas-fired furnace brazing facility had to be built at the Canoga Park plant to accomplish the brazing of this mammoth thrust chamber, which stood 12.5 ft high, and that was just the tubular assembly before attachment of the 16:1 skirt. The injector face had 31 concentric copper rings with LOX and RP-1 distributed in alternate rings. We will be discussing more about the injector design that evolved during engine development.

There was nothing radical about the design of the turbopump other than its pumping capacity and power density. The two compact centrifugal pumps had the capability to empty a large swimming pool in just 10 seconds. Its two-stage impulse turbine had to generate up to 60,000 hp, more than all of the power delivered by all of the giant gas turbines that powered ocean liners like the *Queen Mary*—pretty impressive statistics. The design was kept simple by employing just a single shaft, with the two 3-ft diam turbine discs on one end and the LOX pump on the other. The fuel pump, sandwiched between, was fed by two opposite side inlets in a configuration known as the "baby pants."[5] Each centrifugal pump had an axial flow inducer to prevent cavitation of the main impeller. The shaft was supported by three bearings, a roller bearing at the turbine end plus two ball bearings, all fuel cooled. There were in all nine dynamic seals to ensure no mixing of LOX, fuel, and turbine gases. The real challenge was the fact that the seals and bearings had to operate at 1500°F at one end of the shaft and −300°F at the other end.

Solving Combustion Instability

As expected from the short-duration King Kong testing, combustion instability was an immediate problem. The use of a rough combustion cutoff (RCC) device became routine. Little was known about the instability phenomenon at that time, and the accepted approach was basically just trial and error testing of injector patterns and starting sequences. This appeared to be working. An injector pattern of like impingement, LOX-on-LOX and fuel-on-fuel, along with a rather slow start sequence gave good results. On 26 May 1962 an F-1 engine was fired at Edwards at its full 1.5 million lb of thrust for its full required flight duration of 2.5 min.[6] Time for celebration, but the celebration did not last long.

One month later on 28 June 1962, engine 108, during its third test and after firing smoothly for 105 s, blew up. Bob Biggs describes it as follows, "A combustion disturbance occurred with unprecedented ferocity. The initial shock was violent enough to cause the high-pressure fuel ducts to rupture, and the resulting LOX-rich operation completely destroyed the engine. No safety cutoff device could have prevented the total engine loss and, before the problem was finally solved, two more engines would be lost."[7]

To attack the problem, MSFC set up a combustion stability ad hoc committee chaired by Jerry Thomson. At Rocketdyne a special team was led by Paul Castenholz, Dan Klute, and Bob Fontaine. Bob Levine led the support from the Research Group. Combustion experts from all over the country were consulted.[8] It took all of two years, but the understanding of the combustion instability phenomenon was greatly advanced. Using high-frequency pressure sensors with oscilloscope readouts, the dominant acoustic mode was determined to be the "racetrack" mode, with pressure waves sweeping around the periphery of the injector face. The trick was then to efficiently break up that racetrack. Many detailed changes to the injector were tried, but the most efficient solution was the addition of baffles, which were internally cooled by fuel that was then injected into the chamber so that there was no loss in combustion efficiency. Scaling up the baffles used on the Atlas engines had not worked initially, but the F-1 injector was eventually broken up into 13 zones with a pattern of radial and circular baffles, breaking up the major acoustic resonance zones. This produced very stable combustion. Small bombs were detonated in the chamber with the criterion that this deliberately induced instability had to be damped out in less than a 10th of a second. Several engines were put through repeated bomb tests to demonstrate firm stability.

Achieving Reliability

Combustion instability was not the only major hurdle. The LOX pump experienced a number of failures, and because pure oxygen burns just about anything and everything, the end result was invariably an explosive fire that

Paul Castenholz   Dan Klute   Bob Fontaine   Bob Levine

consumed all of the evidence. The normal suspect areas were bearings and seals, or else a rubbing of the impeller. Those areas were checked in detail, but meanwhile there were four pump explosions on component test stands and five during engine tests. Then the team got lucky. On 18 November 1963 during an engine test at MSFC, the oxygen pump failed but did not catch fire. On teardown of the pump after the test, one of the six vanes designed to guide the flow from the inlet into the impeller was found to have broken loose. Instead of rubbing and causing a fire, it had wedged itself in the impeller flow passage. The smoking gun had been found. These vanes curved the flow and therefore had high pressure on one side and lower suction pressure on the other, creating a strong structural load. If anything caused cavitation, the suction side pressure dropped even lower, and the load exceeded the strength of the vane. Once the vanes were beefed up, the problem disappeared.

Minor problems like small cracks and erosion and leaks (there were a good number of these with the listing requiring a thick document[9]) were attacked with the same determination and attention to detail as the major problems. This was one of the great strengths of Rocketdyne. The design and development groups known inhouse as "Big Engineering" under the leadership of Paul Vogt, Bill Brennan, and Norm Reuel had a fantastic capacity for detail, willing and able to spend countless hours concentrating on the finest detail of a component. Bob Biggs, a member of the F-1 team, said it well, "I was privileged to work with a most persistent and dedicated group of people. Long hours and weekends were not avoided, and, most important, a certain tenacity for details permeated the organization from top to bottom. Problems were to be understood, and 'snow jobs' were promptly dispatched. It was an organization to take great pride in."[10] This was what it took to make an engine of better than 99% reliability, fit for human passengers.

To say that engine testing was extensive would be an understatement. Almost 3000 engine tests were conducted with 40% of them of durations equal to or greater than the F-1's design duration of 2.5 min. The reliability demonstration program included 336 consecutive successful tests, which statistically equated to a reliability of 99% at a confidence level of 96.6%.[11] The F-1 completed its flight rating tests on 16 December 1964 and was declared by NASA to be ready for cluster firings and actual flights. Production commenced, and all manned flight-rated engines were delivered on or ahead of contractual schedules. In total, 98 F-1 engines were delivered to NASA.

Von Braun had conducted hot-firing tests of the Saturn I booster stage with its eight H-1 engines at Huntsville, and he had received a number of complaints from the local citizens about the loud noise. They had not experienced anything yet. The test firing at MSFC of the five F-1 engine cluster at 7.5 million pounds of thrust was a whole new experience for the citizens of Huntsville. Fortunately for NASA the neighbors had become fully aware of the economic importance of MSFC to the city of Huntsville, and they learned to live with their dishes rattling in their cupboards and even a few broken windows. Later development and production testing was done on new test stands at ERB and, starting in 1966, at NASA's Mississippi Test Facility (MTF). This facility is located in the Mississippi Delta region 40 miles from the city of New Orleans. Large engines and even complete launch vehicle stages can be brought to this site by barge. The test stands are quite distinct from those at Santa Susana and ERB. Rather than having work platforms supported by a truss structure of steel beams, at MTF the decks are mounted to massive concrete structures that house offices, parts supply rooms, and a complete array of functions to support the test operations.[12]

In 1974 MTF was renamed the National Space Technology Laboratories (NSTL). In 1988 President Ronald Reagan renamed it the Stennis Space Center (SSC) in honor of Mississippi Senator John Stennis. Noise was not a problem in this isolated location, which was fortuitous, as the enormous energy released into the atmosphere by the combined force of five F-1 engines "shook the heavens" and provided a unique experience to all observers. Even the veteran test engineers and technicians never got used to the experience. When it came to Saturn V launches at the Kennedy Space Center, the public, dignitaries, and even NASA officials were going to be in for a new experience.

The working relationship between Rocketdyne and the people at MSFC up to this point was very interesting. The MSFC crew was very much "hands on," quite capable of assembling and testing hardware. Rocketdyne delivered engines to them, and the Marshall personnel did the handling, maintenance, and testing, with technical advice and assistance from Vince Wheelock, Rocketdyne Field Manager in residence at MSFC, and a few Rocketdyne field engineering personnel. Rocketdyne also had at MSFC a very capable spare parts management and warehousing function under the capable leadership of Norm Dingilian. Their principal management interface with MSFC was Lee Belew and his Engine Project Office. There was only one on-site Rocketdyne technician, Murray Myers, who had supported MSFC for many years with engine specialized maintenance. Rocketdyne was quite comfortable with this arrangement. They respected the ability of the MSFC engineers and technicians and enjoyed working with capable hardware people rather than "paper pushers."[13] This was soon to change when the Saturn V production engines started to be delivered to various engine installation and test sites, where less skilled vehicle contractor personnel were performing engine maintenance and checkout. By 1967 higher management at NASA decided that it wanted more 'accountability', i.e., if something went wrong with a Rocketdyne engine, they wanted to hold Rocketdyne fully responsible, not some vehicle contractor technician. From that point on there were substantial numbers of Rocketdyne field operations and quality control people at the Saturn engine installation, test, and launch sites. This practice carried over to the Space Shuttle Main Engine Program.

In Chapter 7 we discussed George Mueller's big gamble in specifying an all-up configuration of all three stages for the initial launch of the Saturn V on 9 November 1967. Bob Biggs was at Kennedy Space Center for that launch and still remembers the experience well, "Rather than sit at a console and watch data, Dave Aldrich and I stood at the large window of the newly completed Firing Room Two to watch the spectacular event. Apollo Saturn 501, bigger than the Statue of Liberty, lifted off and climbed skyward. The sound of the five F-1 engines was unbelievably awesome, even from three miles away. It sounded more violent than the space shuttle and contained a low frequency component that tuned in and caused a vibration of my rib cage, and even a lower frequency that caused displacement of the large window in front of us. Nearby, the normally poised Walter Cronkite summarily lost his composure and took several distinct steps backward in his mobile broadcast booth."[14] Remember, this was three miles from the launch pad and inside a sturdy building.

Walter Cronkite was not the only one startled by the launch. Ted Benham, Rocketdyne's senior project engineer on the F-1s, saw the Saturn V start to tip as it lifted off the pad and had a moment of panic as he thought it was going to fall over and explode. He was later informed that the Saturn V was designed to translate slightly sideways at liftoff to provide a safer clearance with the launch tower.[15]

The Historic Apollo Missions to the Moon

On just its third launch on 21 December 1968, the F-1- and J-2-powered Saturn V carried its first human crew of Frank Borman, Jim Lovell, and Bill Anders into Earth orbit and beyond. George Mueller, George Low, and the senior directors of the Apollo program had taken another bold step when they approved sending this mission, Apollo 8, all the way to the Moon, where it performed 10 orbits of the moon and sent back to Earth the astronauts' famous Christmas message to all Earthlings. The most lasting impact of that mission was a picture of the limb of the Moon with the tiny blue planet Earth in the background. It really drove home like never before the fragile nature of our small habitat in the vast space of the cosmos.

To call the decision to send Apollo 8 all the way to the Moon 'bold' is really an understatement. Suppose it had failed and lost the crew. After the experience with investigations and hearings following the shuttle *Challenger* and *Columbia* failures, can you imagine the grilling Mueller and Low would have faced? Going to the Moon on the very first manned flight of a Saturn V??!! And after just two test flights?? And after a J-2 engine shutdown prematurely on the preceding test flight of a Saturn V?? George Low was an extremely competent and meticulous engineer who knew every detail of the Apollo system, and when he was confident, that was enough for Mueller to give the go-ahead.

After two more successful Apollo flights to demonstrate spacecraft functioning came one of the greatest moments in the entire history of mankind when on 20 July 1969 the Apollo 11 mission succeeded in landing Neil Armstrong and Buzz Aldrin in the Sea of Tranquility area on the surface of the moon, while Mike Collins supported them from lunar orbit. Anyone fortunate enough (and old enough) to be watching on television that night will vividly remember the thrill of seeing the television pictures they sent back of the Lunar Module and implanted American flag sitting proudly on that ancient dusty surface of our neighbor in space. The lunar rocks they brought back, some several billion years old, are still being analyzed for what they can tell us about the origin of our solar system.

The Saturn Vs performed beautifully in all of the manned Apollo launches, sending the Apollo 12 crew of Charles Conrad, Richard Gordon, and Alan Bean to a 19 November 1969 touchdown in the Ocean of Storms area of the Moon. Intentionally targeted near the robot spacecraft Surveyor III, which had landed on 20 April 1967, the Apollo 12 crew was able to detach and return with the Surveyor's television camera, which showed surprisingly little damage from its long exposure in a space environment. While the 11 April 1970 launch of Apollo 13 was another Saturn V success,[16] a failure in the Service Module spacecraft oxygen system gravely endangered the crew of Jim Lovell, John Sweigert, and Fred Haise. The entire world was united as never before (or since) in hoping and praying for their safe return. The movie '*Apollo 13*' is no exaggeration of the ingenuity and heroic effort it took to get them safely back home to Earth. Apollo 14 was a much smoother ride for Alan Shepard, Stuart Roosa, and Edgar Mitchell. The irrepressible Shepard made a unique historical mark with the first golf shot made on the Moon. The Apollo 15 crew of David Scott, Alfred Worden, and James Irwin had the luxury of a Lunar Roving Vehicle to greatly speed their excursions to lunar surface features of interest. The Apollo 16 crew of John Young, Ken Mattingly, and Charles Duke also benefitted from the speed and range of a Lunar Roving Vehicle. Besides, it looked like great fun to spin "doughnut" circles and send the lunar dust flying in great plumes.

While lunar science was not the primary reason for the Apollo program, each mission certainly returned contributions to scientists' knowledge of the Moon. Instrument packages were left behind to transmit years of data to Earth, and the astronauts were trained to recognize rocks of interest and where to collect soil core samples. However, the science community kept pushing NASA to send a professional geologist to the Moon. The scientists finally got their wish on the last mission, Apollo 17, when a trained geologist, Harrison "Jack" Schmitt, was selected to join the crew of Eugene "Gene" Cernan and Ronald Evans. Jack collected samples of the most ancient and interesting lunar rocks of all of the Apollo missions. (After his career as an astronaut, Jack followed in the steps of John Glenn and was elected a U.S. Senator.) Apollo had achieved its main objective of dramatically demonstrating to the world the technical leadership of the United States. Science had never been its primary goal, and the program came to a conclusion when the Apollo 17 crew landed back on Earth on 19 December 1972.

It is a little known fact that there are five Rocketdyne J-2 engines on the Moon. NASA elected to have the spent S-IVB third stages of the Saturn V launch vehicles for Apollo missions 13–17 impact on the Moon to provide an input to the scientific seismometers already implanted on the lunar surface. Each stage had a single J-2, and so Rocketdyne has the remains of five engines as mementos on the Moon.

Throughout the manned spaceflight program, the Rocketdyne-powered Saturn IBs and Saturn Vs had performed beautifully. Russian engineers had developed many fine high-performance rocket engines, but Sam Iacobellis reports that the Russians repeatedly told him that they considered the F-1, due to its simplicity and 100% reliability, the 'finest rocket engine ever built.'

In total, Rocketdyne delivered 98 of the F-1 engines to NASA. The 13 launches of the Saturn V with its five-engine S-IC stage were powered by 65 of these engines. The unused spare engines became real collector's items—standing 20 ft tall they made most impressive monuments. Indeed one standing in Huntsville has been officially registered as a National Historical Monument. Another stands proudly at the main entrance to the Rocketdyne plant in Canoga Park. Others are featured in museums like the National Air & Space Museum (NASM) in Washington, D.C.

Wernher von Braun visits Rocketdyne's production facility; left to right: Ross Clark, Rocketdyne Plant Manager; Wernher von Braun, NASA MSFC Center Director; Eberhard Rees, NASA MSFC Center, Deputy Director, Technical; and Dave Aldrich, Rocketdyne F-1 Engine Program Manager.

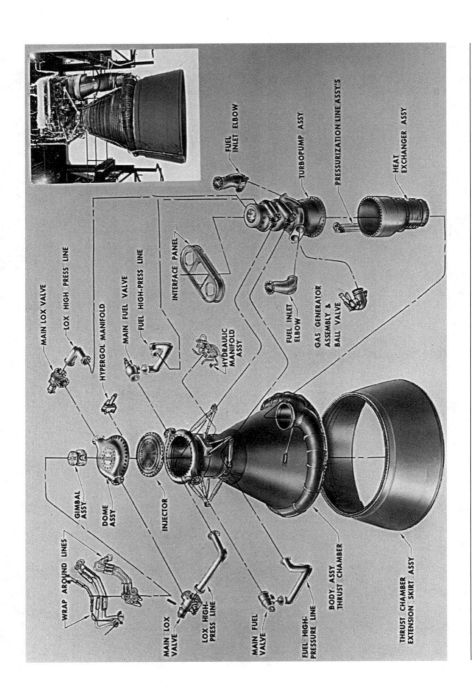

F-1 engine major components breakdown.

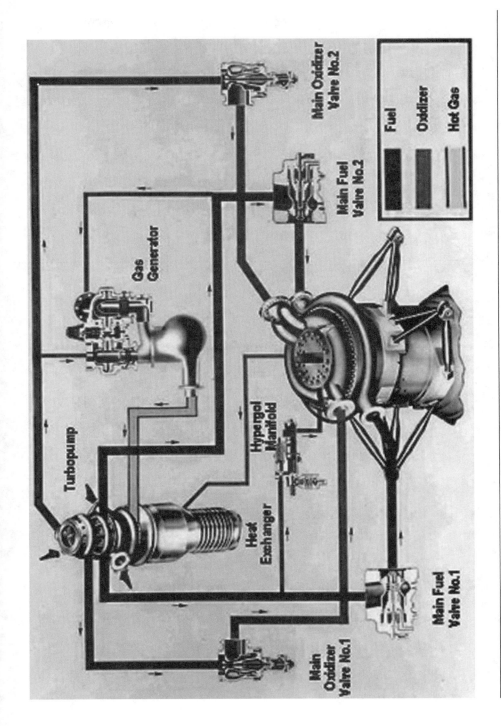

F-1 engine schematic.

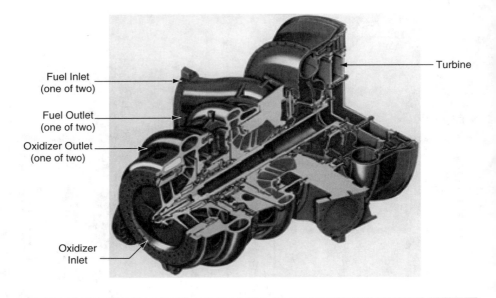

F-1 engine Mark 10 turbopump cutaway.

F-1 engine main injector.

Thrust, pounds		
	Sea level	1,522,000
	Vacuum	1,748,200
Specific impulse, seconds		
	Sea level	265.4
	Vacuum	304.1
Chamber Pressure (psia)		1,125
Engine mixture ratio		2.27
Mission duration, seconds		165
Qualification Life		
	Starts	20
	Duration, seconds	2,250
Weight, pounds		18,616
Production Deliveries		98

F-1 engine characteristics.

F-1 engine installation at Edwards Field Laboratory (EFL), 1962.

F-1 engine hot-fire closeup, 1962.

F-1 engine hot-fire test from distance, 1961.

Thermal protection system (TPS) installed on F-1 single engine, 1965. As an open aft engine compartment on the S-IC stage, the engines each required a TPS like above.

S-IC-1 stage rollout at Marshall Space Flight Center (MSFC), 1968. The first two S-IC flight stages were fabricated and hot-fire tested at MSFC by a NASA-Boeing team. Subsequent S-IC stages were fabricated by Boeing at the Michoud Assembly Facility (MAF) and tested by Boeing at the Mississippi Test Facility (MTF) with the exception being S-IC-3, which was tested at MSFC. The stages were then barged to Kennedy Space Center (KSC) for launch.

S-IC-T five-engine hot-fire cluster test at Marshall Space Flight Center (MSFC), 1965.

Fourth flight stage of the Apollo/Saturn V space booster (S-IC-4) with five Rocketdyne F-1 engines is installed in the S-IC stage cluster hot-fire stand at Mississippi Test Facility (MTF), 1967.

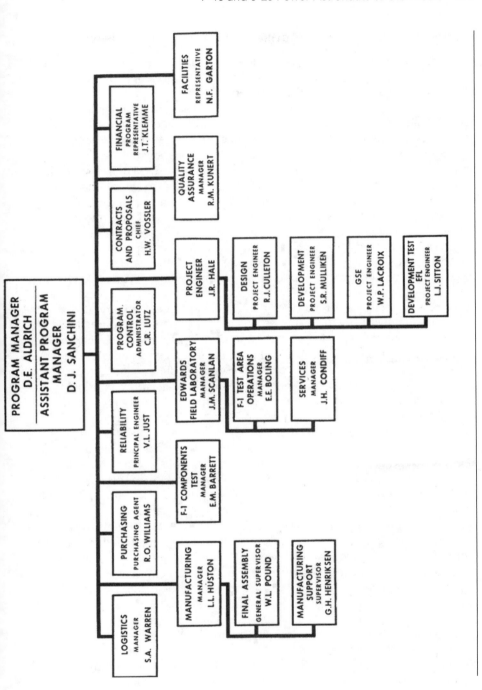

Rocketdyne F-1 program organization chart.

Rocketdyne Rescues the Apollo Ascent Engine

Certainly Rocketdyners are justly proud of their contributions to the spectacular success of the Apollo program and the first human landings on the Moon. The record should show that their contribution extended well beyond just the F-1 and J-2 engines. If one includes all of the Rocketdyne-provided thrusters (both liquid and solid propellant) for attitude and velocity control and propellant settling, the total number of rocket reaction devices they provided to each of the Apollo lunar missions ranged from 26 to 34. One small but vital system they are especially proud of was the Ascent Engine for the Lunar Module (LM). This was a very simple engine that the LM contractor, Grumman, had subcontracted to Bell Aerospace. The Ascent Engine had a vacuum thrust of 3500 lb and was designed for a single continuous burn of up to seven minutes to lift the Apollo astronauts off the surface of the moon and carry them to their rendezvous in lunar orbit with the Apollo Command/Service Modules. Absolute reliability was required if the astronauts were ever to get back home. Accordingly the system was kept extremely simple—just pressurized tanks, no pumps, and an uncooled ablative thrust chamber. The storable propellants were nitrogen tetroxide (NTO) with a 50/50 mix of hydrazine and unsymmetrical dimethyl hydrazine (UDMH). Because of the hypergolic nature of this propellant combination coupled with the modest thrust level, stable combustion was pretty much assured. Surprisingly, the engine showed persistent combustion instability, but with a unique characteristic. It did not increase in amplitude to destructive levels, but neither did it damp out, rather continuing at a constant amplitude for the duration of the firing. All hands agreed that this was not an acceptable anomaly for Apollo.

Various combinations of injector patterns were tried by Bell without success. Injector baffles had worked on the Atlas and F-1 engines at Rocketdyne, and these were tried by Bell, but they caused unacceptable erosion at the throat of the ablative nozzle. In solving the F-1 instability problem, Paul Castenholz at Rocketdyne and Jerry Thompson at MSFC had developed a small bomb that could be detonated in the combustion chamber. The chamber was judged to be stable if the bomb-induced oscillations damped out within 400 ms. Bell started employing the bomb tests, but the Ascent Engine kept flunking the tests. Combustion experts from all over were called in, but the problem was getting desperate. The entire test schedule for the LM was being adversely affected. Apollo 8 was supposed to carry the LM, but it had to be deleted as not ready to fly. At NASA George Mueller, Gen. Samuel "Sam" Phillips (Apollo Program Director), and George Low (Apollo Spacecraft Director) kept a list of "show stoppers," and the Ascent Engine moved to the top of the list.[17]

Meanwhile in August 1967 Rocketdyne had been contracted for a relatively low-budget but fast-moving backup effort under the direction of Small Engines manager Steve Domokos and key team members Cliff Hauenstein, Mike Yost, Ram Martinez, Ed Houston, Tim Harmon, Ed Shuster, Paul Lewis, and Terry Trinan. The team wanted state-of-the-art equipment for both injector electron beam welding and for precision electrical discharge drilling of injector orifices. Sam Hoffman without hesitation approved the order using company funds.

The team judged that Bell's triplet injector pattern had impingement too far from the injector face, allowing too much unburned propellant in the chamber that could power combustion instability. The Domokos team settled on a mixed doublet pattern (fuel on oxidizer) with impingement close to the injector face. In just a few months the team had developed a stable injector that also incorporated acoustic cavities around the perimeter and injector baffles. Perhaps an overkill, but this thrust chamber had to be stable with a wide margin of confidence, as demonstrated by quick recovery in repeated chamber bomb tests. The chamber problem appeared to be solved, but the rest of the engine system still had bugs to work out before it would be ready for production. In May 1968, in a bold move requiring much intercompany cooperation, NASA, Grumman, and Bell agreed to move the Bell engine to Santa Susana and have Rocketdyne integrate their injector. In a crash effort, within just a month the Rocketdyne modified engine had passed 53 consecutive bomb tests without instability. The entire engine was completely qualified by August 1968, and a production engine was delivered from Rocketdyne in time for the 3 March 1969 launch of Apollo 9, the first test flight of the LM. The production engine for the historic Apollo 11 lunar landing mission was delivered in May 1969, just in time for the launch in July.

The Ascent Engine crew finally got to see their product in action on the Apollo 15 mission, whose Lunar Roving Vehicle had a television camera and transmitter for sending moving pictures of the liftoff back to Earth in real time. The camera did not have very good resolution, but the LM was seen clear enough to record any mishap at ignition of the Ascent Engine. Of the Rocketdyne viewers, typically nervous when humans were riding on their rockets, Steve Domokos and his team were the most relaxed—they knew the margin of reliability that they had built into that engine. Sure enough, ignition was smooth, the exhaust scattered a spray of lunar dust, and the ascent portion of

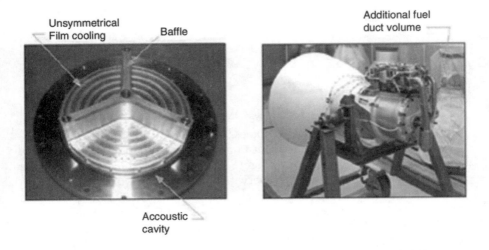

Lunar Excursion Module (LEM) Ascent Engine Rocketdyne injector design.

182 Rocketdyne: Powering Humans into Space

NASA JSC personnel (left) and Steve Domokos (RIGHT) inspecting LEM Ascent Engine at Rocketdyne.

the LM was soon out of view as it headed for lunar orbit and its rendezvous with the Command and Service Modules.

The Rocketdyne-integrated Ascent Engines performed flawlessly on all six of the Apollo landings on the Moon. Steve Domokos still glows when he recounts that story, and he has every right to be proud of what his team accomplished. He worked on and managed many projects during his years since joining Rocketdyne in 1951, but he labels the LM Ascent Engine as his favorite, proclaiming, "The program was fast, crisp, and most of all extremely successful."[18]

Saturn V/Apollo spacecraft launch.

CHAPTER 9

The Fierce Competition for the Space Shuttle Main Engine

Wernher von Braun had set future goals for human space flight in his 1952–1954 articles in *Colliers* magazine. His initial destination was the Moon, but his long-range goal was to land humans on the surface of Mars. A necessary step toward that goal was to have a large manned space station in Earth orbit to develop the life support systems, radiation shielding, exercise programs, simulated gravity, and other techniques that would allow humans to survive the long trips to and from Mars in good health. The station would also provide a base in orbit for assembly of the spacecraft that would go to Mars. Such a space station would need regular supply visits, which would be prohibitively expensive using expendable launch vehicles, and so von Braun proposed a fully reusable launch vehicle with a winged upper stage that would fly back to the launch area to be refueled for the next launch. His proposed steps in chronological order were as follows: 1) develop a reusable launch vehicle, then 2) assemble and utilize a permanently manned space station, and finally 3) assemble and launch the spacecraft to carry humans to land on Mars, initially for exploration and eventually to establish a Mars base. When NASA was established in October 1958, it adopted those goals, at least in spirit, and whether formally stated or not has pursued them ever since.

NASA Sets High Performance Goals for Space Shuttle Engines

In the late 1960s NASA began design studies and technology development for the reusable launch vehicle, which was named the Space Shuttle by George Mueller, NASA's head of manned space flight. Studies quickly revealed the sensitivity of the designs to the performance of the liquid-propellant rocket engines that would boost the shuttle all the way from takeoff into orbit. Just a few percent increase in the rocket engine performance (specific impulse) could substantially reduce the gross weight of the vehicle. NASA, with input from Air Force propulsion people, started to support technology development to raise engine specific impulse. In a 1969 NASA competition for the pre-Phase B preliminary design of an advanced hyrogen/oxygen engine, Paul Vogt's people dominated the Rocketdyne proposal team and submitted a very conservative proposal based on just upgrading the existing J-2 engine with better pump bearings and seals. Although the procurement was out of MSFC, the NASA proposal evaluation team was composed mainly of advanced engine technology people supported by NASA's Office of Advanced Research and Technology (OART). They were interested in advanced technology, not detail design upgrades. Rocketdyne came in a distant third to Pratt & Whitney and Aerojet and was fortunate to be included in the later Phase B competition. This experience strongly indicated that NASA, or at least NASA Headquarters, wanted new high-performance technology for the planned Shuttle engines.

Hydrogen/oxygen was already well established as the propellant combination of choice for high performance, certainly for at least all upper stages and probably for all stages. The Pratt & Whitney RL-10 engine for the Centaur stage and the Rocketdyne J-2 engine for the Saturn IB and Saturn V upper stages had well established the readiness of those propellants for the Shuttle.

The next most effective engine parameter for increasing the specific impulse is the expansion ratio of the engine nozzle. The further you can expand the exhaust gases of the engine, the greater is the gas velocity and accompanying momentum change, and hence the greater the specific impulse. In the vacuum of high altitude, this simply means expanding to the highest nozzle expansion ratio possible within weight and volume limits. The J-2 engine's nozzle had a relatively high expansion ratio of 27.5:1, appropriate for its use on upper stages. The much smaller RL-10 went even higher at 40:1.

For takeoff at sea level, the tradeoffs get more complex. If the exhaust is expanded beyond the sea-level ambient pressure, it first experiences overexpansion to less than ambient pressure, subtracting from the net thrust. Any further expansion causes the flow to separate from the walls of the nozzle. If the flow separation is not perfectly symmetrical, then serious side forces can be generated. Thus booster engines are designed with low expansion ratios that are appropriate at sea level but have relatively poor performance at altitude.

One way to get booster engines to higher expansion ratios is to increase the engine's chamber pressure, which allows expanding the exhaust gas further before reaching ambient pressure. Thus while the H-1 engine had a chamber pressure of 700 psia and an area ratio of 8:1, the later F-1 engine was able to go to a higher chamber pressure of 982 psia, and so it could incorporate an increased expansion ratio of 16:1. One could further increase the chamber pressure, but that makes cooling of the thrust chamber more difficult and makes the design of the turbopumps much more challenging. In addition, the power required to drive the pumps to higher pressures becomes so high that it would be terribly inefficient to dump the turbine exhaust gases overboard or even into the nozzle. To maintain good engine efficiency means incorporating much more complex engine system topping cycles that inject the turbine exhaust into the main combustion chamber. Such topping cycles require very high pump discharge pressures, a real challenge to the state of the art. The Rocketdyne Advanced Design group was confident that it had a better solution through novel nozzle design.

Developing the Highly Advanced Aerospike Concept

After developing analytical techniques for the design of shortened bell nozzles, as covered in Chapter 5, the Rocketdyne Advanced Design group continued to pursue the potential of the method of characteristics (MOC). As a graphical technique the MOC projected expansion waves emanating from the nozzle throat. Where each wave hit the nozzle wall a curvature could be designed to generate a compression wave of equal or slightly greater pressure. The stronger the expansion wave, the stronger could be the compression wave and the sharper the nozzle wall bending, leading to a shorter nozzle. The very strongest expansion would be all the way to ambient pressure. This led to the idea of putting a sharp-edged plug with a large blunt base into the throat of a bell nozzle. The plug would be large enough so that the nozzle exhaust flow would be annular with a hollow core at or near ambient pressure. Thus the expansion around the plug's edge in the throat would be all the way to ambient pressure, producing very strong expansion waves that could be matched where they intersected the nozzle wall by very sharp curvature. The resulting nozzles, dubbed 'expansion-deflection' nozzles, or E-D, were extremely short.

Verified by wind-tunnel tests and then hot firings, the expansion-deflection concept was patented in 1959. (Robert S. Kraemer was listed as the inventing team leader.) It is interesting that while the Air Force was the principal supporter of Rocketdyne's technology advancement efforts, this new pioneering in nozzles was largely supported under a contract from the Navy's Bureau of Aeronautics (BuAer) headquartered in Baltimore.

While it looked great on paper, the E-D design would be difficult to incorporate into large engine designs. Just the cooling of the plug would be a challenge, but creative minds did not stop there. Someone asked, "Why not turn it inside out?" Imagine running a knife down the center line of a cross section of the E-D nozzle, cutting it in half. Then put the two halves together, but back to back, so that the exit walls of the nozzle meet at a

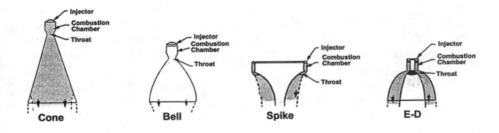

Rocket engine nozzle shapes.

point, forming a "spike," with the two halves of the plug throat now on the periphery. An engine with such a nozzle would have an annular combustion chamber (or annular arrangement of small combustion chambers) exhausting into an annular throat, where the gas would expand immediately to its surrounding ambient pressure. That meant the resulting strong expansion waves could be balanced with a strongly curved nozzle wall, resulting in a shortened spike.

Nozzle shortening was just one beneficial feature of the spike configuration. Because it expanded the exhaust immediately at the throat to the surrounding ambient pressure, it automatically adjusted expansion as it went up in altitude. It closely approached optimum expansion all the way through the boost phase of a launch vehicle up to its full expansion ratio, which could be made very large—truly a remarkable performance advance.

Refinements followed. How about shortening the nozzle still further by cutting off the tip end of the sharp spike? That would leave a blunt base that would see low pressure and hence create negative thrust. Why not duct the turbine exhaust there to increase the pressure and gain thrust from the turbine exhaust instead of just dumping it overboard? The resulting engine configuration with a sawed-off spike has been aptly described as resembling "an upside-down volcano."[1] Because the turbine exhaust flow aerodynamically replaces the cutoff tip of the spike, the configuration was named the "Aerospike." Verified by wind-tunnel tests and hot firings all the way up to 250,000 lb of thrust, this Aerospike design worked like a charm. It was determined that the length of the Aerospike could be cut down to as little as 1/5 the length of a bell nozzle thrust chamber.

Here was the answer to a rocket designer's prayer—an engine that could operate at modest chamber pressure with a simple gas generator power cycle but still provide very close to perfect expansion all the way from the equivalent of modest expansion ratios at takeoff to very high expansion ratios at altitude. How often in life do you find a perfect solution like that?

Then here came NASA asking for the highest possible specific impulse at both sea level and altitude from engines for the proposed space shuttle. Under the leadership of Sam Iacobellis, who had succeeded Bob Kraemer as head of Advanced Design, Rocketdyne planned to propose the super advanced Aerospike and accelerated their test program to demonstrate performance at increasingly higher thrust levels. Sam got major support for the Aerospike effort with the reassignment of some top engineers from Vogt's "big engineering" organization to the Aerospike team, including the experienced and highly respected Wilbur "Willie" Wilhelm. Progress was rapid, with strong inputs from Advanced Design members Hank Wieseneck, Bob Sobin, and Bill Wagner, and outstanding work by development engineer Paul Fuller.[2] They found that development could be done on just a narrow slice of the Aerospike, no more than two inches wide, demonstrating more than 99% combustion efficiency in a combustion zone only 2.5 in. long and producing the predicted expansion down the spike. A full-spike thrust chamber was then just an assembly of these multiple slices.

Ed Monteath (left) and Sam Iacobellis discussing design features of annular Aerospike engine.

NASA Leans Toward High-Pressure Staged Combustion

Technical progress was going extremely well, but enthusiasm within NASA was mixed. Von Braun expressed strong support, as did Max Faget, the chief designer of the space shuttle, but not all of the engineers at MSFC were convinced that the technology was ready. George Mueller at NASA Headquarters also had strong reservations. Ideally the Aerospike would utilize the entire base of the launch vehicle stage as a surface to expand against, thereby saving vehicle weight. It would differentially throttle its annular ring of combustion chambers to provide thrust vector control. However, Mueller was concerned that this would require a close partnership between the engine contractor and the airframe contractor. He had planned to have those be two separate procurements and contracts. Rocketdyne had always believed in hardware demonstrations rather than just talk and so responded by demonstrating a gimbaled version of the Aerospike that could be installed like any other engine without being incorporated into the airframe structure—pretty convincing.

The other approach to high specific impulse was to go to very high chamber pressure, at least as high as 3000 psia, so that a large area ratio nozzle

Annular Aerospike hot-fire test at the Nevada Field Laboratory (NFL), Reno, Nevada. Annular Aerospikes were extensively tested at NFL with thrust levels ranging from 4000 lb to 250,000 lb and using a variety of propellants.

could be employed without over-expanding the exhaust even at sea level. However, it takes a lot of energy to pump the propellants to very high pressure, and so the design has to find a way to minimize any energy loss. Pratt & Whitney had demonstrated with their upper-stage RL-10 engine for Centaur that hydrogen warmed to a gaseous state during passage through the thrust chamber cooling jacket could drive the turbopumps and then be burned in the main combustion chamber for good efficiency. This system was known as an 'expander cycle.' It had not proved to be easy to develop. Because any change in hydrogen pressure affected the entire system, it took extensive computer analysis and simulations to finally learn how to stabilize and calibrate the engine. The final development cost for the RL-10 went to quadruple the original Pratt & Whitney estimate.[3]

A key to making the expander cycle work was that the RL-10 had a modest chamber pressure of only 300 psia. The expander cycle could not generate enough energy to power the pumps to very high discharge pressures, like well over 7000 psia for a chamber pressure of 3000 psia. To pump propellants for a 3000-psia chamber pressure, a portion of the warm hydrogen would have to be burned fuel rich with some of the oxygen in a gas generator. To avoid the lost energy in the gas generator exhaust products, they would have to be kept at high enough pressure to burn in the main combustion chamber. With this system the gas generator was better described as a 'preburner'. Because of the two stages of combustion, first in the preburner and then in the main chamber, the system was called a 'staged combustion cycle'.

NASA's Del Tischler, head of propulsion in OART and a very likeable and intelligent engineer, along with his rocket specialist Hank Burlage, surely must have appreciated the merits of the Aerospike concept. However, Rocketdyne was amply supporting its development with both government and corporate funding, and so Tischler decided to advance the high chamber pressure approach and put out for bid a contract to test an experimental thrust chamber at a chamber pressure of 3000 psia. This was to be a very small chamber, and most Rocketdyne engineers thought it too small to prove much of anything. In a very internally controversial decision, Rocketdyne decided not to bid on that program, which went to Pratt & Whitney. That would help to consolidate a Pratt & Whitney leadership position in high chamber pressure. Things took an ominous turn for Rocketdyne when, in a speech on 23 October 1969, George Mueller described the engines for the planned Space Shuttle as "high pressure, staged combustion rocket engines."[4]

The initial reaction in the Advanced Design group at Rocketdyne was that surely this position of Mueller's would change. In selecting a power cycle for the Shuttle's main engine, NASA would have to balance the factors of performance, cost, and risk. Surely the Aerospike was superior to a high-pressure engine on every one of those parameters.

Thus the stage was set for by far the most important competition in the history of liquid rocketry in the United States. Future engines would either stay at relatively low pressures and a simple gas generator cycle, but with dramatically new nozzle designs, or else go to very high pressures with much more complex power cycles. The future prospects, and even the existence, of the three leading American rocket engine development companies, Aerojet, Pratt & Whitney, and Rocketdyne, hinged on the outcome of this crucial battle. All three companies were hurting for rocket work, and the Shuttle engine development was the only big development contract in sight.

The Phase B Shootout with Pratt & Whitney

In 1969 NASA awarded three contracts to Aerojet, Pratt & Whitney and Rocketdyne for Phase B preliminary design studies of a new space shuttle main engine (SSME). Aerojet lacked experience with hydrogen/oxygen engines and was never in serious contention. The contracts said that study results and any developments under NASA contracts were the property of the U.S. Government, and so all results were openly disclosed in a series of presentations during 1970 to

NASA representatives from MSFC and Headquarters. Sam Iacobellis did his usual masterful job of presenting the status and merits of the Rocketdyne Aerospike design, and Dick Mulready was also effective in presenting the impressive subsystem development work that Pratt & Whitney had done toward a staged cycle high-pressure design, including their Air Force funded work aimed at a 250,000-lb thrust staged combustion engine labeled the XLR-129. Thus Pratt & Whitney knew of Rocketdyne's Aerospike design, and Rocketdyne in turn was fully knowledgeable of the Pratt & Whitney high-pressure staged combustion engine design. Rocketdyne engineers had achieved their past success by developing reliable engines utilizing the relatively simple gas generator cycle. They did not want to get into the complexity and extreme technical difficulties of a high-pressure staged combustion cycle. While they assembled a team (which included future Rocketdyne General Manager Byron Wood) to study alternate high chamber pressure system designs, they kept selling the merits of an Aerospike engine.

This was a critical time, with the future of Rocketdyne strongly dependent on winning the SSME competition. It was no time for management shakeups,

Byron Wood

but there was a disruption in 1969. When NAA had merged with Rockwell-Standard in September 1967 to create North American Rockwell Corporation (NAR), the NAA product divisions including Rocketdyne had been placed in the NAR Aerospace and Systems Group under John Moore as President. Moore was a very successful manager who had built the Autonetics Division into a very successful operation. He did not know Rocketdyne very well and wanted to put his own people in control. In September 1969 he sent Jay Wethe to be his representative at Rocketdyne. Wethe immediately demanded the best office at the Canoga Park plant, which resulted in Sam Hoffman being booted from his office. Sam was a proud man, and he had reason to be, and he did not take this well. Neither did all of his managers and dedicated personnel. Sam departed ("retired") a few months later in May 1970, as did Tom Meyers, who had come back from Astrodyne to oversee plant operations for Hoffman. Wethe never did gain the level of respect the Rocketdyners had for Hoffman, and he left Canoga Park in April 1971. John Moore left North American Rockwell for a high-level position at McDonnell Douglas.

Sam Hoffman (left), Rocketdyne President, Dr. Wernher von Braun (center), NASA MSFC Space Flight Center Director, and Dr. Eberhard Rees (right), NASA MSFC Space Flight Center Deputy Director, Technical.

Reluctant Decision to Propose Staged Combustion Cycle

Sam Hoffman was succeeded as President of Rocketdyne in 1970 by the popular Bill Brennan, who, just like Hoffman, grew to depend heavily on Doug Hege for the winning of new business. Iacobellis reported to Doug, and the two communicated daily (even on weekends) on the marketing campaign for the SSME. Both of them came to the reluctant conclusion that Pratt & Whitney's high chamber pressure approach was going to win. Von Braun may have been sold on the Aerospike, but he was not going to be on the source selection team at NASA. In fact, he had been transferred to NASA Headquarters in February 1970, and so he was no longer the Director of MSFC, which was now led by von Braun's long-time deputy Eberhard Rees.

Rocketdyne was going to miss von Braun in many ways. Members of management who dealt with him are unanimous in their respect for him as a great organizer and motivator of people. Like Sam Hoffman, he was smart enough to leave the detailed technical decisions to his technical experts. With him out of the SSME picture, and with some members of the selection board coming from OART, Doug and Sam became convinced that the vote would go to high chamber pressure, and not only high chamber pressure but also the staged combustion cycle.

Brennan approved the quick assembly of two 100-man teams, one led by Paul Vogt to concentrate on a design evolved from the J-2 engine, and the other led by Sam Iacobellis with Willie Wilhelm and Dom Sanchini on a high chamber pressure staged combustion design. The designs were reviewed at the end of 30 days, and Brennan accepted the assessment of Doug Hege and made the decision to concentrate on the staged combustion design as being what the customer wanted. Whether you believe that the customer is right or wrong, in the end after you have made your case, you need to give him what he wants. Sam Hoffman had always preached that philosophy.

Bill Brennan

The dilemma was that some at Rocketdyne, including Chief Engineer Paul Vogt, doubted that the staged cycle engine could ever be developed. Moreover, there was a widespread belief at Canoga Park that Pratt & Whitney did not have the resources to develop such a complex engine. If NASA were to get the engine it wanted, Rocketdyners sincerely believed that they would have to be the ones to deliver it. Doug Hege summarized that position, "It's my opinion if Pratt & Whitney had won they would have had to hire most of Rocketdyne's staff because they didn't have the depth of engineers in the disciplines necessary to develop their engine."[5] This was a prejudiced opinion, without doubt, but as we will see later, Pratt & Whitney management agreed that there was some validity to Hege's judgment.

Under the oversight of Matt Ek and Sam Iacobellis, Dom Sanchini was given the vital role of leading the proposal preparation team, which included standouts Willie Wilhelm (a key person on engine design), Roger Burry, Hank Wieseneck, Vern Larson, Bill Wagner, Don Fulton, Frank Larry, Paul Combs, Spencer Clapp, Ron Goe, Vince Wheelock (stressed ease of field servicing), and many others too numerous to list. They did an outstanding job, completing a highly detailed and comprehensive proposal that included 100 thick volumes. Paul Castenholz had done an outstanding job as Rocketdyne's program manager for the development of the highly successful hydrogen/oxygen J-2 engine and was held in high esteem by the engineers at MSFC, and so he was named in the management plan as Rocketdyne's program manager on the SSME.

At this juncture a small incident occurred at MSFC that illustrates the tension of the competition for the SSME contract. Bob Biggs had been assigned to MSFC as a point of contact on the SSME, and he had made friends with a number of the people there. One evening he was in a restaurant having beer and pizza with one of his MSFC friends, just "shooting the bull." He was relating that he had had to get help from his friend Dom Sanchini in convincing Paul Castenholz to put him on the proposed SSME team. Unknown to them, a senior Rocketdyne executive at a meeting at MSFC on another program was in an adjoining booth just around a corner and overheard part of the conversation, which he mistakenly thought was between two MSFC people. He immediately phoned Canoga Park, asked for Bill Brennan, and told him that MSFC people were having a difficult time working with Castenholz. Panic! Was there a need to quickly change the proposal to name someone other than Paul as program manager? Fortunately a few phone calls straightened things out, and Castenholz stayed in the proposal. It is amusing to look back on that now, but with proposal tensions running high, it was a serious matter at the time.

Castenholz and his team now entered deeply into the preparations for the staged combustion engine. One of the key decisions was whether to have one preburner or two, with a separate preburner for each turbopump. Certainly one was simpler, but two gave more flexibility in startup and shutdown sequencing. Aside from technical tradeoffs, Castenholz argued that the Rocketdyne design must be visibly distinct from the Pratt & Whitney single preburner design. He had a significant influence on the decision to go with dual preburners.[6] The resulting flexibility in sequencing proved to be vital during the later painstaking development of engine start sequences. Moreover, the dual preburners allowed close control of mixture ratio and thrust while eliminating the need for any hot gas throttle valve.

Crash Program to Demonstrate Prototype

Castenholz, always a man of action, strongly advocated that there was no way they were going to beat Pratt & Whitney with just a paper proposal. Rocketdyne had always demonstrated its new concepts with real hardware, not just paper studies. Paul proposed a crash program to build a prototype engine, with at least the preburner power head assembly and thrust chamber, and fire it at the SSME design chamber pressure and thrust level. Bill Brennan met with Bob Anderson, Chairman of North American Rockwell and, following in the tradition of the old North American leaders Kindelberger and Atwood, Anderson approved that very substantial investment in the future of Rocketdyne.

Castenholz quickly assembled a team and was off and running. A key to the design had to be how the chamber walls were cooled. They knew that Pratt &

Whitney had developed an intricate wall material with labyrinth passages for transpiration cooling of the combustion chamber walls. The hydrogen used for this surface cooling would not be fully burned and would subtract from the engine specific impulse, negating some if not all of the gains from the high chamber pressure. Moreover, this extra hydrogen would lower the overall oxidizer-to-fuel mixture ratio, penalizing the shuttle with larger hydrogen tanks. Any advantage of the staged combustion cycle was seriously eroded, if not eliminated.

North American had always had superb metallurgists on their staff, including the brilliant Dr. Zuech, and for the Aerospike they had already come up with an alloy of copper and zirconium that retained the excellent thermal conductivity of copper but had a much higher melting temperature, promising to transfer heat through the walls without a big temperature gain and thereby maintaining wall integrity. They named it NARloy-Z, with the NAR in tribute to North American Rockwell. Clearly any chance that Rocketdyne had of winning the SSME contract would be greatly enhanced if NARloy-Z really worked. It did work, as the Advanced Design group led by Floyd Bennett demonstrated regenerative cooling of small thrust chambers at up to 5000 psia chamber pressure in firing tests at NAR's remote Nevada test site.

A new fabrication technique had to be developed to make a thrust chamber with NARloy. Ingots of NARloy-Z had to be spun into the contour of the thrust chamber and then more than 200 channels machined to fine tolerances for the coolant passages. After fine grinding, sanding, and electrochemical etching, the chamber was installed in a large tank for electroforming of the outer surface of the thrust chamber. All of these procedures had to be worked out and tested before the SSME proposals were due in April 1971.

Iacobellis and Hege were right in their assessment of how NASA was leaning. The NASA Space Shuttle Program Requirements Document dated 1 July 1970 had specified "400,000 pound sea level thrust bell-type engines." While not final, pending Phase B results, it was certainly an omen. Sure enough, on 1 March 1971, NASA's official request for proposals on the SSME ruled out the Aerospike by specifying that the engine had to utilize a bell nozzle. Moreover, the engine requirements were so specified that only a high chamber pressure staged combustion cycle design could satisfy them. The pressure was really on Paul Castenholz to hot test a high-pressure thrust chamber before the proposal presentation. The proposals had to be submitted by 21 April 1971. During those grueling days Paul rarely left the office, sleeping on a cot in the Rocketdyne medical office.

The Rocketdyne prototype SSME thrust chamber and power head with preburners were assembled on a test stand by a crew under the direction of Ted Benham, with strong support from Dom Sanchini, at the company's Nevada Field Laboratory, a remote test site about 20 miles from Reno. Hot-fire testing was started in early 1971. Time did not permit the development of an automatic ignition detector as a part of the start sequence. Instead, a pillbox was partially buried about 50 yd downstream of the thrust chamber, which was mounted in a horizontal position. Ted Benham positioned himself in the pillbox, looking out through a thick windshield and with a signal button in his hand. Only after observing the glow of preburner ignition would Benham press the button that would allow the ignition sequence for the main chamber to continue. After progressive tests on the preburners, the first test was initiated to take the main chamber into partial mainstage. Even with gaseous hydrogen coming into the chamber, Benham says he was just able to make out the flickering glow from the preburners, and he pressed his go-ahead button. Ted says the resulting fierce exhaust stream immersing the pillbox, along with all of the dust, sand, and pebbles it carried, scared him half to death.[7]

While the test demonstrated a good start and stable combustion, and there was no hardware degradation, the duration was only 0.35 s of main chamber firing, and the chamber pressure reached only 2084 psia, short of the magic goal of 3000. Moreover, the thrust chamber cooling had not quite reached thermal equilibrium to demonstrate the success of NARloy-Z. Bill Brennan was understandably concerned about risking the chance of a failure by conducting another test before the submittal of the SSME proposals. Over objections, Castenholz charged right ahead anyhow, and on a snowy Reno day on 12 February1971, his team conducted a second mainstage test firing that stabilized at a chamber pressure of 3172 psia for 0.45 s. That may not seem long, but it was sufficient for the recorders to verify that thermal equilibrium and excellent specific impulse had been achieved.

NFL large-engine test stand where SSME proposal engine was tested, 1970.

SSME proposal engine installed in NFL large-engine test stand, 1971.

SSME proposal engine hot-fire test, 1971.

SSME Phase B proposal pressure-fed engine design team; back row, left to right: Don Brown, Bob Unden, Paul Dennies, and Les Nave; middle row, left to right: Bill Mower, Marsh Pollack, Ross Barnsdale, and Floyd Bennett; front row, left to right: Ron Cook, Lyle Fakler, Harry Dodson, Jake Reidyk, Irv Eberle, Wayne Munyon, and Frank Kubic.

The test engine assembly was then mounted in a large moving van along with poster charts of the instrumentation readouts as well as videos of the tests and trucked back to MSFC in Huntsville to show to one and all connected with the SSME competition. Paul gave a super presentation with multiple-screen projections—what attendee Bob Biggs calls "the briefing that saved Rocketdyne." Eberhard Rees, the Director at MSFC, was especially impressed, which did not hurt at all. Biggs says he heard Rees turn to one of his people and say, "Now I know it can be done."[8] The van was then driven to NASA Headquarters in Washington, D.C., where it drew a steady stream of interested NASA managers and specialists. On 13 July 1971, NASA announced that it had selected Rocketdyne to design, develop, and manufacture the Space Shuttle Main Engines.

SSME proposal engine mockup, volumes, and test engine.

SSME team, 1971.

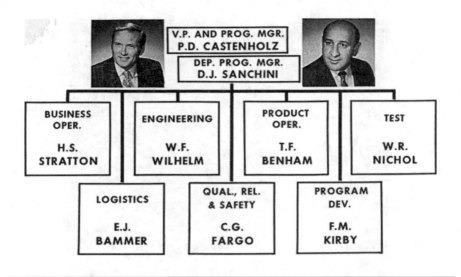

Rocketdyne SSME program organization chart.

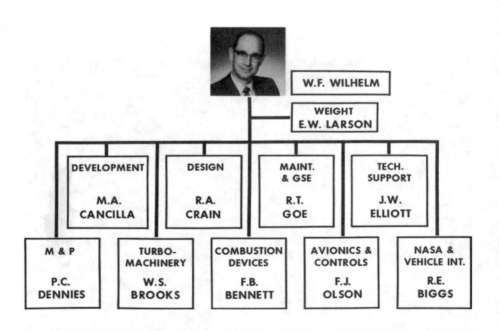

Rocketdyne SSME Engineering organization chart.

The Fierce Competition for the Space Shuttle Main Engine 203

Needless to say, there was a great celebration at Rocketdyne. Employment had fallen off from over 20,000 in the peak Saturn V launch vehicle days to just 3600 by 1970, and the SSME was the only engine development. prospect in the foreseeable future. The campaign to win the contract had been one more demonstration of the outstanding capabilities of Rocketdyners. Take Sam Iacobellis, for example. In 1973 Sam was promoted to the position of President of the Atomics International (AI) Division of NAR, then head of the Energy Group of NAR, which included both AI and Rocketdyne, followed in 1986 to the position of President and COO of Rockwell's Aerospace Operations, and then in 1994 to Deputy Chairman of all of Rockwell. He was a pretty good sample of the talent that came out of Rocketdyne.

Sam Iacobellis

The Shock at Pratt & Whitney

Not everyone in NASA was thrilled that a staged combustion engine design had been selected. Max Faget, the chief designer of the shuttle at the Johnson Space Center (JSC), was dismayed at the complexity of the engine and the high pump pressures required. Iacobellis and crew presented him with several realistic designs that would produce the performance gains of high chamber pressure (up to 5000 psia) but without the topping cycle features so that pump discharge pressures could be held to only 6000 psia. Specific impulse would be close to that of the staged combustion design. As expected, Faget had no success in selling this at NASA Headquarters.

Across the country in Florida, Pratt & Whitney had been working for 10 years on the technology for staged combustion engines, including the test of impressive high-pressure turbopumps and other subsystems. To say they were disappointed at not winning the SSME contract would be an understatement. They lodged a formal protest with the General Accounting Office (GAO) alleging that NASA had not run a fair competition. Rocketdyne engaged lawyers to match the Pratt & Whitney lawyers, but all of this caused delays until the GAO rejected the Pratt & Whitney protest on 31 March 1972.

There is little doubt that the hot-firing test was key to Rocketdyne's win, but the proposal evaluation scoring was very close. A strong factor had to be a long history of North American's commitment to Rocketdyne. The people at MSFC knew that Rocketdyne and North American Rockwell would expend every possible effort to develop the SSME. While Dick Mulready of Pratt & Whitney was extremely bitter over their loss, and promoted their unsuccessful protest of the award to the GAO, he did admit that he never had the full backing of upper management at Pratt & Whitney and its parent United Aircraft Corporation. He wrote that "when push came to shove, the vital interest simply was not there. Whenever the question arose of a possible separation of the assets, including people, for any new venture into a stand-alone organization to do any new business, the jet engine managers always voted to keep the effort as an appendage of their group. In his interview taped for the fortieth anniversary of Pratt & Whitney FRDC, Dick Coar [Mulready's boss] reflected that if we had won the Space Shuttle engine development work, it would have been in competition with the F100 [jet engine program] and we could not have done both."[9]

The Complexity of the SSME

The 100-volume Rocketdyne proposal package had encompassed design, development, and management detail, but now the design had to be made to work. The staged combustion SSME marked an entirely new and markedly different direction for Rocketdyne engines. Up until now all of its large engines had used the gas generator cycle pioneered by Robert Goddard and Wernher von Braun with his German V-2. Rocketdyne had placed special emphasis on simplification, as exemplified by the beautifully simple X-1 and H-1 engines. Even the massive F-1 engines and the high-performance hydrogen/oxygen J-2 engines were straightforward designs using the bipropellant gas generator cycle. The staged combustion cycle design for the SSME was much more complex and challenging. Not only were the chamber pressure and heat transfer rates much higher, but also the pump discharge pressures had to be up to two and one-half times the main combustion chamber pressure. Pressures approaching 8000 psia were going to severely challenge pump seal designs. Power densities through the pump turbines were higher than had ever been achieved before. The control stability margins were so narrow that developing a smooth start sequence was challenging to the degree that it could never have been achieved before the development of sophisticated computer models. To aid in controlling the start, Rocketdyne had prudently elected to have separate preburners for each of the propellant pumps, as opposed to the Pratt & Whitney design with only one preburner.

Paul Vogt, the conservative Chief Engineer at Rocketdyne, was so distressed at the complexity of the SSME that he wrote a letter in protest to Eberhard Rees at MSFC, saying that Rocketdyne management had "prostituted" itself in proposing a terribly complex and difficult staged combustion design rather than a straightforward uprated J-2 design against their better judgment and solely in order to win the competition. Vogt's judgment was that the staged combustion design might even be impossible to develop at any price. Now one must respect Vogt's concern, and he had good reasons to question the staged combustion design and to debate its adoption within Rocketdyne, but there was no excusing his going to the customer and undermining his top management's decision. Bill Brennan really had no choice but to dismiss his friend and former boss from Rocketdyne for this "mutiny" against management. Vogt was transferred to the Space Division with no meaningful job, and so he retired and was replaced as Chief Engineer by Matt Ek. It was really too bad that Paul should exit in this manner—he had made such great contributions to the development of Rocketdyne's series of exceptionally reliable rocket engines.

One of Paul Vogt's concerns was the inherent complexity of the staged combustion SSME design. Perhaps the best way to illustrate that complexity is to reproduce the following engine description given by Bob Biggs, the SSME Chief Project Engineer. In this excerpt, the initials LP and HP stand for low pressure and high pressure, TP is turbopump, F is fuel, O is oxidizer, PB is preburner, and MCC is main combustion chamber.

"The LH2 enters the engine at the low pressure fuel turbopump (LPFTP) inlet at a pressure of 30 psia and is increased in pressure by the 15,000 RPM turboinducer to over 250 psia. This pressure is required to prevent cavitation of the high pressure fuel turbopump (HPFTP). The three-stage centrifugal pump (a departure from the J-2s seven-stage axial pump to give more margin against stalling), operating at 35,000 RPM, further increases the pressure to over 6,000 psia. The LH2 is then divided into three separate flow paths. Approximately 80 percent of the fuel flows to the two preburners; half of this, however, is used to cool the thrust chamber nozzle and then mixed with the other half prior to entering the preburners. The remaining 20 percent of the fuel is used in the major component cooling circuit. The LH2 is first routed to the MCC where it provides coolant for the main combustion process by flowing through 390 milled slots in the copper alloy combustor. Having been converted to an ambient temperature gas by the MCC, the fuel is then routed to the LPFTP where it is used as the power source for the partial admission single stage impulse turbine which drives the LPFTP. A small portion (0.7 pounds per second) of this gas is then used by the Space Shuttle to pressurize the main hydrogen tank while the rest of it is used to cool the major hot gas system structure (hot gas manifold) and, finally the main injector baffles and faceplates before being consumed in the MCC.

[This is just the fuel system. The oxidizer side is still to come.]

"The LOX enters the engine at the low pressure oxidizer turbopump (LPOTP) inlet at a pressure of 100 psia and is increased in pressure by the 5,000 RPM turboinducer to over 400 psia. This pressure is required to prevent cavitation of the high pressure oxidizer turbopump (HPOTP). The dual inlet single stage centrifugal main impeller, operating at almost 30,000 RPM, further increases the pressure to about 4,500 psia. Most of the LOX is then routed through the main oxidizer valve to the coaxial element main injector of the MCC. A small amount of LOX (1.2 pounds per second) is routed through an engine-mounted heat exchanger and conditioned for use as the pressurant gas for the Space Shuttle main oxidizer tank. The remainder of the LOX is ducted back into a smaller boost impeller on the same shaft to increase the pressure to as much as 8,000 psia. This provides enough pressure to allow the use of throttle valves to control the LOX flow rate into the two preburners. Thrust control is achieved by closed loop throttling of the oxidizer preburner (OPB) side, and mixture ratio control is accomplished by closed loop control of the fuel preburner (FPB) side. The throttle valves are controlled by an engine-mounted computer known as the main engine controller (MEC). A built-in recirculation flow path provides power for the six stage axial flow hydraulic turbine which drives the LPOTP. A LOX flow rate of approximately 180 pounds per second is supplied from the discharge side of the main impeller; and, after passing through the turbine, this LOX is mixed with the discharge flow of the LPOTP and thereby returned to the HPOTP inlet.

The two preburners produce a hydrogen-rich steam that is used to power the two high pressure turbines that drive the HPFTP and the HPOTP. Combustion of these gases is completed in the MCC."[10]

All of these subsystem elements are packed around the powerhead of the engine, filling up every available cubic centimeter of space. It is a wonder that the engine can be assembled and serviced. To an admirer of the beautiful simplicity of the H-1 engine, the SSME is a nightmare. However, the customer wanted maximum specific impulse via high chamber pressure and staged combustion, and so that is what the customer got.

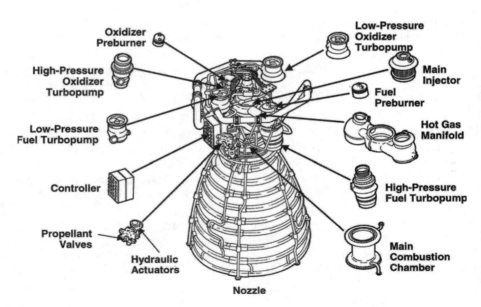

SSME major components.

The Fierce Competition for the Space Shuttle Main Engine 207

Power Cycles of Pump-Fed Liquid-Propellant Engines

Gas Generator Cycle
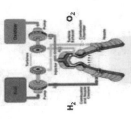
Advantage - Simplicity
Disadvantage - Low performance
Rocketdyne

Tap-Off Cycle
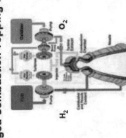
Advantage - Simplicity
Disadvantage - Low performance

Staged-Combustion Topping Cycle

Advantage - High performance
Disadvantage - Complexity, cost

Expander Cycle
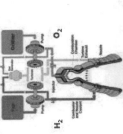
Advantage - Simplicity, good performance
Disadvantage - Limited Pc, weak start

 J-2S
 RS-68
 SSME
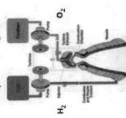 RL-10

Power cycles of pump-fed liquid-propellant engines.

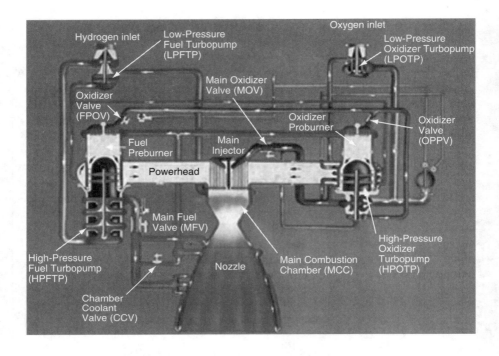

SSME propellant flow schematic.

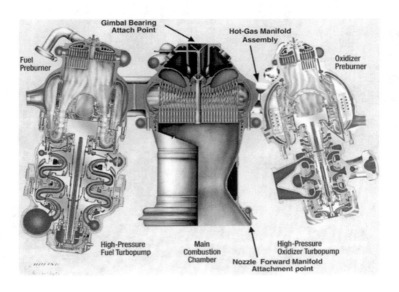

SSME Powerhead component Arrangement, Block II.

Senior NASA and Rocketdyne Shuttle program personnel viewing SSME Preburner at Rocketdyne, left to right: Owen Morris, NASA Integration Project Manager at JSC; Bill Brennan, Rocketdyne President; Mike Malkin, NASA Shuttle Program Director, NASA Headquarters, Washington D.C.; Bob Thompson, NASA Shuttle Project Manager, JSC; Frank Stewart, NASA SSME Project Manager, MSFC; and Paul Castenholz, Rocketdyne SSME Vice President and Program Manager.

The Problems Begin

As might be expected, problems were not long in surfacing. Executing some design changes to achieve weight reduction, coupled with necessary structural improvements, caused the fabrication of major components and subsystems to fall behind schedule. In any event the test facilities were not ready. The normal course of development would be to conduct development tests of the subsystems before progressing to engine testing. However, the complex nature of the SSME required component and subsystem test facilities of unprecedented complexity. Two subsystem test facilities in the Coca area of Santa Susana were modified to try to duplicate all of the system transients predicted by sophisticated computer programs. Each facility required about 2000 valves, many servo-controlled, to produce the predicted system dynamics. The program slipped behind schedule.

Meanwhile, the relations with MSFC top management had taken a rather sharp downturn. Von Braun had practically adopted Rocketdyne as his personal rocket engine development team. Castenholz and the MSFC engineers had the greatest respect for each other. Paul especially liked the readiness of the Marshall people to come out to Canoga Park and help work out solutions to technical problems.[11] When von Braun moved to NASA Headquarters in 1970, he was replaced by his compatriot Eberhard Rees, who may not have been quite as warm and charismatic as the incomparable von Braun, but relations were still very good. Rees's successor, Rocco Petrone, was a different person.

Rocco was a "muscle man" both physically and in management style. He was quick to respond to problems with strong action. With any threatened schedule slip or cost increase, he leaned very hard on all people involved. He shuffled his people at MSFC, replacing Frank Stewart with J. R. Thompson as his new SSME Project Manager. He climbed all over Paul Castenholz, who did not take it well. Paul worked well with Frank Stewart and the MSFC engineers, had an incomparable record of success as a dynamic engine program manager, and he was confident that he could move the SSME development as fast as anyone. He acknowledged that the budget was strained—inflation was running higher than projected (nothing he could do about that), and fabrication costs were beginning to run above estimates. He did not blame Petrone for being concerned about costs and schedule slips, nor for his applying strong-arm pressure, but Rocco did not seem to want to hear specifics about the problems and what Paul was doing to correct them. Paul himself was about as aggressive as one could be and still maintain morale and lead an effective team effort. He considered Petrone to be far too aggressive, saying Petrone "didn't want to hear what I had to say, did not want to hear that we would need more money. He did not want to hear my analysis of why. . . . I think he was a brute-force bully, and that doesn't get the job done."[12]

In such a situation, the common corporate practice everywhere was (and still is) to replace the program manager and try to start over with a clean slate. In June 1974 Norm Reuel, who had for years overseen all of Rocketdyne's engine development, was appointed to succeed Castenholz, to be followed soon by Dom Sanchini, who had directed the SSME proposal effort. This dictated quite an adjustment for the development team, as Paul and Dom had quite different management styles. Paul was more informal and his meetings were participatory, while Dom's style has been described as being more "autocratic." However, both in their own ways were highly effective.

SSME was the only large project at Rocketdyne, leaving Castenholz with no other suitable project to manage, and so he elected to transfer to the NAR Science Center and then later to NAR Headquarters to work with John Tormey before going to join ex-Rocketdyner Bob Byron as a Division Director for EnviroTech, a Rockwell subsidiary that was developing pollution control systems. Rocketdyne lost one of its all-time leaders.

The head rolling was not to end with Paul. Petrone started chewing hard on Bill Brennan, and so corporate management felt compelled to replace him with Norm Ryker in 1976. That was probably not a great choice, as Norm was very sharp but knew nothing about rocket engines and could not discuss development problems in any detail with the MSFC engineering managers. Fortunately, Sanchini and MSFC's J. R. Thompson got along well. Both were tough-minded and decisive leaders, and there were the inevitable clashes of judgment, but they respected each other and were able to negotiate productive courses of action. Over the tough years ahead, they became good friends.

Dom Sanchini with Space Shuttle Main Engine documentation.

Under Sanchini the vital role of chief project engineer was assigned to Willie Wilhelm, who had played a key role in the proposed design of the engine. In 1976 Willie was succeeded by Bob Biggs, who reported to Paul Fuller, the chief program engineer. The manager of engine design was Bob Crain, who put a lot of emphasis on designing for ease of ground servicing between launches.

Spectacular Failures

Turbopump testing in the Coca area at Santa Susana began several months behind schedule with the low-pressure pumps in May 1975, progressing to the high-pressure turbopumps three months later. Then during a test of the oxidizer subsystem (LPOTP, HPOTP, OPB) on Coca-1A in early 1976, a facility flowmeter failed, causing a major fire that seriously damaged both the components and the facility. Once given an ignition source, a stream of high-pressure liquid oxygen will consume all of the metal around it in the blink of

Norm Ryker

an eye before any shutdown procedure can even be initiated. Rocketdyne engineers wryly referred to this as "hardware-rich combustion." In the area of a failure, there would be nothing left to examine. The engineers had to rely on extensive instrumentation for clues to the problem.

Such fires were not restricted to just the oxidizer subsystem tests. A few months later in 1977 during a test of a fuel subsystem on Coca-1B, some cavitation in a facility throttle valve started erosion that led to another major fire and extensive damage. Even a relatively minor failure somewhere in the fuel supply could reduce fuel flow and result in an oxidizer-rich burnout in the preburner and a resulting major fire. Both of these fires were due to faults in the test facility, not the engine components. More time and hardware were required for development of the complex facilities. The NASA budget was far too tight to provide for all of the needed replacement components, and so the decision was made jointly by MSFC and Rocketdyne management to phase out the Coca area component and subsystem test facilities and concentrate on engine system tests. There was more than a bit of rationalization in the Congressional testimony of the NASA Administrator Robert Frosch when he stated, "We have found that the best and truest test bed for all major components, and especially turbopumps, is the engine itself."[13] A more honest assessment would

Bob Crain

have been that the OMB-allowed budget did not provide the resources to support a normal and logical component test program.

NASA decided that the engine development would be done on the test facilities developed for Saturn engine testing at the Mississippi Test Facility (MTF), later named the National Space Technology Laboratories (NSTL) and then the Stennis Space Center (SSC). The test configuration, known as the integrated subsystem test bed (ISTB), employed flight-design components, except the main engine computer was a rack of ground-mounted electronics, and the thrust chamber nozzle was shortened to an area ratio of 35:1 rather than the flight configuration of 77.5:1. This shortened nozzle would avoid flow separation even when the engine was throttled down as much as 50%. The first test of the ISTB was conducted on 19 May 1975, and full-blown engine development testing was under way.

Painstaking Solutions

Five years of work with computer models had demonstrated how sensitive the staged combustion system was to small changes in timing. A first hurdle had to be purging the engine and getting smooth propellant flow to the three combustors, i.e., the two preburners and the main combustion chamber. The hydrogen flow through the thrust chamber cooling passages caused unavoidable oscillations that essentially set a clock for the rest of the start sequence. The cold liquid hydrogen encountering the warmth of the nozzle hardware expanded rapidly, causing a flow blockage and actual momentary flow reversal. The resulting pulsating fuel flow had a very reproducible frequency of 2 Hz with dips in pressure occurring at approximately 0.25, 0.75, and 1.25 s until the main combustion chamber pressure could be stabilized at 1.5 s. All other start sequence steps had to be timed to hundredths of a second to fit with this oscillation, otherwise the stopping of fuel flow would cause disastrous oxygen-rich combustion in the preburners and main chamber. As an example of the required timing, priming throughout the system had to be achieved at the fuel preburner at 1.4 s, at the main combustion chamber at 1.5 s, and at the oxidizer preburner at 1.6 s. Establishing successful timings and working all of the multitude of possible failure modes out of the start sequence took until late in 1978.

In flight the entire precise sequence and split-second sensing of possible failure modes would have to be done by the main engine controller (MEC), and that was far from being available for testing with the engine. Rocketdyne had perhaps not made the wisest choice in picking the Honeywell division in Clearwater, Florida, for the job of developing that sophisticated controller. It was that subcontractor who had almost caused NASA's Viking Mars Lander mission to miss its launch window in 1975 due to their plant manager's reluctance to invest company funds in the tooling necessary to build the Viking computer's plated wire memory. The NASA Planetary Programs office did not believe in headhunting for scapegoats, but they were compelled to pressure Honeywell top management into replacing the plant manager with someone who would try to fulfill their contract.[14] Tough-minded George Low at NASA Headquarters remembered that well when he noted that Rocketdyne on the SSME "has done a very poor job so far on controlling costs and on controlling its main subcontractor, Honeywell." He went on to note that Honeywell "has done a lousy job on this, just as they have on the Viking computer."[15]

The next big problem encountered in early 1976 was a large-amplitude high-frequency vibration in the high-pressure fuel turbopump. The frequency was at about one-half the pump speed and was immediately recognized as a phenomenon know as subsynchronous whirl. Following their usual pattern on problem solving, Sanchini assembled a team of multidisciplinary experts augmented for this problem by historical research, literature surveys, mathematical models, and consultations with universities and other companies with related knowledge or experience. This problem was solved pretty quickly through a stiffening of the shaft and bearing supports and providing extra dampening at the seals.

Then in 1977 there were two failures of the HPFTP turbine blades just two weeks apart. The two-stage turbine was only 11 in. in diameter but had to spin at a rate of 36,000 rpm and generate a fantastic 75,000 hp, more than the total output of the huge gas turbines on such super ocean liners as the *Titanic* and the *Queen Mary*. Each individual turbine blade was only the size of a small postage stamp, yet had to produce 660 hp. The overall HPFTP was the highest power density rotating device known anywhere in the world today, producing 100 hp per pound of weight. (Can you imagine a drag racer with that kind of output? It would burn its tires to the hub in a fraction of a second.)

The problem team immediately consulted experts throughout government, industry, and academia and started test programs at Rocketdyne, using a machine known as the Whirligig, and at General Electric, TRW, and Ai-Research. The problem was quite quickly identified as a fatigue failure and was solved by improving the mounting and damping on each blade. No change in blade material was required, remaining a directionally solidified casting of a nickel-based super alloy developed by Martin Metals and known as MAR-M-246. Pratt & Whitney was later to comment that a better solution would have been to eliminate the thermal-stress fatigue by going to the hollow blades developed by their Pratt & Whitney jet engine experts. This was probably true, but the damping did the job.

Other problems arose, and solutions were frequently elusive. Spread over the next four years there were four more explosions of the high-pressure oxidizer turbopump. Causes had to be sought in the burned-out remnants. Then there were two burn-throughs of the fuel preburners, a fire in the main oxidizer valve, two separations in the "steer horn" fuel lines to the nozzle (looked like antlers), and a rupture of the main fuel valve housing. Multidisciplinary teams were immediately assembled to probe the causes and find permanent solutions. Ed Larson, Rocketdyne's Director of Design Technology, was a frequent choice to head these special teams, which included members from MSFC and covered the specialties of structural analysis, dynamics, materials, thermodynamics, metallurgy, systems, quality control, data handling, component design, and test planning. Dom Sanchini held daily five o'clock status meetings with the leaders of current problem teams plus key members of the project team, including Bob Biggs (Chief Project Engineer and director of the engine test effort), Paul Fuller (Chief Program Engineer), Jerry Johnson (Associate Program Manager for engine systems), Jim Hale (Associate Program Manager for turbomachinery), and frequently J. R. Thompson (MSFC Project Manager), who was provided with a permanent office right next to Sanchini's and spent as much time in Canoga Park as he did back home in Huntsville.

In this book we will not even try to cover the intense diagnostic work and the detailed design, technique, material, and procedural solutions to all of these problems. For readers interested in the fascinating detective work required to trace down causes and develop reliable solutions, Bob Biggs has written an excellent comprehensive account in two thick papers titled "Space Shuttle Main Engine; The First Ten Years", and its in-work followup, "Space Shuttle Main Engine; The Second Decade."[16,17] His detailed account will really impress the reader with the complex interactions between components that helped make problem solutions so difficult.

Ed Larson

Forced Top Management Turnover

First with Rocco Petrone and continuing with his successor, Bill Lucas, the MSFC Director continued to put extreme pressure on Rocketdyne management. Bill Brennan did not appreciate the strong-arm treatment from Rocco and did not contest his replacement in 1976 by rocket newcomer Norm Ryker. Norm was more accepting of the pressure, but he was soon in disfavor with Bill Lucas and was replaced in 1983 by Rocketdyne veteran Dick Schwartz. Thus MSFC had forced a succession of Rocketdyne presidents from Brennan to Ryker to Schwartz. A pattern of head-hunting was getting established at MSFC after the departure of von Braun and Rees. Petrone believed in strong and decisive action, including changing key assignments, which many would rate as commendable, but the resulting management climate can make people reluctant to reveal any problems in their area of responsibility for fear of being demoted or even fired. A developing scapegoat atmosphere might help explain some of the later apparent communications shortcomings at MSFC and its contractor Thiokol, where a few people seemed to be reluctant to report to their bosses the worrisome seal problems with the shuttle solid-propellant booster motors that doomed the January 1986 *Challenger* launch.

Achieving Reliability

John Yardley, the highly respected manager of the Space Shuttle program at NASA Headquarters, had set a criterion of 65,000 s of accumulated time on SSME system testing as representing a level of development maturity that would qualify the SSME as flight worthy. Drawing from his years of experience at

McDonnell Aircraft, he adopted that number from standard practice in the military aircraft business, where a new fighter would be released for operational use after it had achieved a string of 40 flight tests. Most of the astronauts at that time were former test pilots, and so they had no complaints with Yardley's criterion. Yardley arrived at the magic number of 65,000 by first multiplying 40 with the nominal 520-s SSME burn time of a Shuttle mission: 40 × 520 = 20,800 s. That would be the accumulated time for any one engine, but presumably for extra conservatism Yardley multiplied that by three for the three engines on each Shuttle, to arrive at 62,400, which he rounded out to 65,000 s.[18]

Engine testing moved into longer and longer burn times with fewer and fewer anomalies. The uncooled portion of the thrust chamber nozzle was found to be flexing, but stiffening that was an easy fix. Honeywell finally delivered the compact main engine controller that would have to sense all of the operating conditions throughout the engine, precisely time all the intricate sequences, and provide for safe instantaneous shutdowns in the event of anomalies. With test time rising rapidly, production engine 2004 pushed the accumulated test time over the goal of 65,000 s on 24 March 1980, and the SSME was officially deemed ready for its first flight. By the time of the actual launch a year later, the SSME engine test program had accumulated 110,253 s during 726 hot-firing tests.

The Shuttle Flies

Engines 2005, 2006, and 2007 were assigned for the first flight. Because of changes since their flight acceptance tests, the hot-firing tests were repeated before the engines were installed in the Shuttle orbiter. Then with the complete vehicle assembled on the launch pad at Kennedy space center, solid boosters included, the three-engine SSME cluster was given a 20-s Flight Readiness Firing, which went very smoothly. The solid boosters were of course not ignited. After correcting a minor glitch in the orbiter computer systems, the shuttle was ready for its first launch.

Any rocket launch is an emotional experience, even for observers not associated with the program. For the rocket engineers it is far more intense. They had for years witnessed all-too-frequent fiery explosions of engines on the test stands. For the rest of their lives, especially for the launch of missions with humans aboard, their heart rate would be elevated as the countdown approached T minus zero. With astronauts John Young and Robert Crippen aboard, the final countdown went smoothly, and liftoff of Shuttle *Columbia* on mission STS-1 occurred right on time at 7:00 a.m. on 12 April 1981. Operation of the three SSME engines was flawless. Following the commands of the orbiter guidance and control computers, the engines throttled down from rated power level to 65% to reduce the peak dynamic pressure on the vehicle. After 15 s, past peak "q" forces, the thrust was returned to 100% until the acceleration approached 3 g, at which point the engines were gradually throttled toward 65% to limit the acceleration as propellant was consumed.

Bob Biggs very nicely summed up the remainder of the mission as follows: "The Space Shuttle orbiter Columbia achieved its predicted orbit and remained there for two days. Americans were back in space after an absence of almost six years. Return from orbit occurred on April 14, 1981, with a safe landing at Edwards Air Force Base, California. After the post-flight inspections at the Dryden Flight Research Facility found the engines to be in perfect condition, Columbia was returned to KSC on April 28 to prepare for the next flight. The first reusable spacecraft had been sent to space, safely returned to earth, and was ready to go again. The era of the Space Shuttle had begun."[19] It had taken six years since the start of SSME testing to get to this milestone, but the future looked bright indeed.

The first four Shuttle launches were considered development flights, and all went relatively smoothly. There was some post-flight deterioration in the high-pressure fuel turbopumps that would need investigating, but in general the STS flights were off to a very good start. STS-3 presented an interesting challenge because record-breaking rainfall had flooded the planned landing area on the dry lake at EAFB. Major General James Abrahamson, who had succeeded John Yardley as head of the Space Shuttle program at NASA Headquarters, decided to go ahead with the launch on schedule but switch the landing site to the Northrup Strip at the White Sands Proving Ground in New Mexico, where the designated landing area was a runway carved out of a large barren gypsum bed with no buildings or equipment. That meant that a well-equipped post-landing processing facility had to be assembled in just five days. A small army of over a thousand was assembled from the shuttle contractors, the armed services, and NASA and a small city housing 1555 tons of equipment sprouted on the desert over a weekend. Vince Wheelock, Rocketdyne Director of Field Engineering & Logistics, led the SSME team in performing their part in this small miracle. Fortunately, Vince is a man of great energy and took this all in stride.

The first "operational" Space Shuttle flight was STS-5, launched on 11 November 1982. It was graded as a great success with no significant engine problems. SSMEs have performed successfully in all operational Shuttle missions to date and have continued Rocketdyne's dominant role in propelling humans into space.

Uprating the SSME

It would be nice to end the story of the SSME development right there after STS-1 through STS-5, but there is at least one more chapter to the SSME story that should be told. The original 100% rated power level (RPL) was 470,000 lb of vacuum thrust, and the SSME was certified at that level before the launch of STS-1. However, from the very beginning NASA had wanted a higher level of thrust for a few future missions with heavier payloads and for potential use in future in-flight emergencies. That level was identified as a 109% full power level (FPL) of 512,300 lb of vacuum thrust. After STS-1 development effort was directed to certify the SSME at that level.

Historically Rocketdyne's engine designs had always been relatively conservative so that thrust upratings had come fairly easily—just crank up the thrust, reinforce any weakness that turned up, and certify at the new higher thrust level. (English car designer Colin Chapman had used that approach in developing a series of very successful lightweight Lotus race cars, although his drivers did not especially like the part about "pushing harder to find out what breaks.") This was a simple method, but nothing had ever come easily with the complex SSME, and raising the thrust level a mere 9% proved to be no exception.

Bob Biggs, in the second of his two comprehensive papers on the SSME development, covers the problems in agonizing detail. He summarizes the effort as follows: "Development of the additional nine percent thrust capability was more difficult than anyone had imagined. Seven major failures were experienced by the end of August 1982. Hardware funding was woefully inadequate for the task at hand, causing delays, encouraging re-use of poor performing components, and performing heroic repairs to salvage damaged parts."[20] Major problems included persistent failures of the LOX posts in the main injector. This was finally solved by flow shielding and beefing up the posts by fabricating them from Haynes 188 alloy. LOX posts in the preburners also had to be reinforced. Synchronous and subsynchronous vibration reappeared in the high-pressure oxidizer turbopump. The high-pressure fuel turbopump proved especially troublesome with failure of an endcap and threaded stud known as the "Kaiser hat" in addition to repeated fatigue failures of the turbine blades.

Design improvements were made, and by April 1982 two engines were halfway through their FPL certification cycles. Bob Biggs described the next test:

"On April 7, 1982, Engine 2013 test 901-364 was planned as a 500-second FPL flight mission simulation test. At 362 seconds the engine was destroyed by a failure which originated in the HPFTP turbine. The engine was burned and ripped out of the test stand. It was found 100 ft. away from the base of the test stand, in the spillway built for the flame bucket coolant water. The HPFTP was separated from the engine, by about 60 ft. The HPFTP first stage turbine disk (with no blades) was 100 ft. away from the HPFTP. Beyond that were other engine components, including half the engine gimbal bearing. The other half of the gimbal was still attached to the test stand, as well as the two low pressure turbopumps. The ducting had been ripped apart." [21]

The SSME had a cruel way of letting the engineers know that their latest design improvements were not quite adequate.

Even after the SSME was finally certified at the 109% FPL, continuing problems finally dictated that it be decertified at that level and recertified at a new 104% FPL. The two-year launch hiatus to fix the solid motors following the *Challenger* tragedy in January 1986 allowed time for more SSME improvements. On the problem of fatigue failures of the turbine blades, a strong contributing factor was the thermal stress generated by having the blades first at cryogenic temperatures as the pumps were primed and then subjecting them to hot gases from the preburners, causing high thermal stress within the solid blades. Pratt & Whitney's turbojet people had developed hollow blades that had proved to have much longer lifetimes under thermal gradient conditions. It was not too astonishing when NASA gave Pratt & Whitney contracts to design and develop new high-pressure SSME turbopumps that would have longer lifetimes (the Rocketdyne turbopumps required major inspections and servicing after each flight). Pratt & Whitney was able to achieve their objectives, albeit at the expense of several hundred pounds of added weight. It was painful for Rocketdyne when they had to assemble these Pratt & Whitney units into their SSME engines, but they could not really argue with NASA's seeking a second source. They accepted that it was good national policy to try to keep a minimum of two companies in the United States capable of designing large liquid-propellant rocket engines. Happily, at the time of this writing, the two companies are working well with each other.

If one took a poll among Rocketdyne engineers asking which of their engines was their personal favorite, there would probably not be a single vote cast for the SSME. Looking back on the tortuous SSME development, many at Rocketdyne would tell NASA, "We told you so. You should have gone with our Aerospike design and its much simpler low-pressure gas generator cycle." This is probably correct, but still there has to be a very strong sense of pride at successfully developing such a technically challenging engine.

However, life is full of humbling experiences. One of these for Rocketdyne was to find out later that Russian engineers had developed a whole series of successful large engines using the staged combustion cycle. They flew a staged combustion engine on the upper stage of their Molniya launch vehicle as early as 1960.[22] Most of their staged combustion engines utilized LOX and kerosene and ran the preburners oxidizer-rich. Why did they find it so straightforward to utilize staged combustion? George Sutton, who has made an in-depth study of Russian engines, notes that some of their engines took a very long time to complete development. For example, the RD-0120 engine for the Energiya launch vehicle, the only staged combustion engine they have flown using LOX/hydrogen, took 16 years to develop.[23,24] Thus it was not all that easy for them either.

NASA had set some very challenging requirements for the SSME. Were the Russian specifications as challenging? The SSME had to start with just propellant tank pressure, with no start system to get the turbopumps spinning like on the J-2 engine. Then NASA specified an extremely high chamber pressure to achieve a high specific impulse all of the way through the vehicle boost phase. This dictated the use of the staged combustion cycle, which in turn drove pump discharge pressures and turbine powers still higher. However, the Russian RD-0120 also had a high chamber pressure of 3162 psia. Did the use of hydrogen vs kerosene make starting transients more difficult? Bob Biggs thinks so, writing that "hydrogen is a significant complication compared to RP-1, especially during start, but we overcame that in about a year. A major complication to the SSME was caused by the original design requirement to throttle mixture ratio from 5.5 to 6.5 and thrust from 50% to 109%. This required that the main oxidizer pump and the main fuel pump be controlled separately, and was the primary driver for the engine system layout. Two turbopumps, each driven by its own turbine, powered from its own preburner. The turbopumps were very low inertia [if a pump experienced cavitation it would accelerate to destructive overspeed in less than a tenth of a second], the turbines were very powerful, and the two systems had to be held in balance for start, mainstage and shutdown. We got through this one also. The major cause of SSME failures was maximum performance with minimum weight—everything was designed to the limit."[25] Bob does admit that Rocketdyne was not perfect, with incidents of "parts not made to print, processes improperly applied, environment not well understood, and bad judgment in test planning."[26] Overall the Rocketdyne engineers concur with Biggs when he says, "I still think the SSME was a great success story against huge odds and I doubt that any other group could have completed it as well."[27]

We can hope that further discussion between Russian and American rocket engineers will add to our understanding of their staged combustion development. Meanwhile, to the chagrin of Rocketdyne, the American Atlas launch vehicles, long powered by Rocketdyne engines, starting with the Atlas III series have begun to use staged combustion Russian RD-180 engines.

The energy released during the launch of a Space Shuttle is enormous, and the worry of a launch failure has always been with us, even before the 1986 catastrophic failure of a solid motor on the *Challenger*. Until the breakup of *Columbia* on 1 February 2003, it was seductively easy to concentrate concerns on the launch phase and to discount the slim margins of safety during reentry. Because he landed safely, it is easy to forget John Glenn's scary moments during reentry of his Mercury capsule in 1962. In his own words, "The first time I went up in space, I had two things happen. One, the automatic control system went out and I had to control the spacecraft manually. And the second thing was a signal that came down to two ground stations that the heat shield was loose. So NASA decided that instead of firing the retrorocket pack that slows you down up there, and throwing it away in space and reentering with a clean heat shield, I would re-enter with the retrorocket pack strapped onto the body of the spacecraft. The theory was that it would hold the heat shield in place. So I made a very spectacular re-entry because that retro pack was burning off during re-entry and there were big flaming chunks of it flying past the window. I couldn't be certain whether it was the heat shield or the retro pack. All I could do was to keep on working [manually steering the capsule]".[28]

Glenn was characteristically not one to make a big event of that scary reentry, and so the greater concern continued to be focused on the launch phase. After the *Columbia* failure, he summed it up as follows: "We've never had a disaster on re-entry before. The launch has always been more worrisome because you've got thousands of tons of fuel ready to blow up if something goes wrong, so that's always looked at as a time of more hazard than any other."[29] That the system of probably greatest concern, the complex and highly stressed SSME, has so far performed without failure is a rewarding tribute to the dedication of all of the Rocketdyne people who performed its design, development, manufacture, and servicing.

Despite ambitious plans, NASA's current budget continues to look too lean to readily develop a new generation of lower cost launch vehicles, whether reusable or expendable. Although strictly required only through transportation of the planned large modules to the International Space Station, the space shuttles may be in service for quite a few years to come. The SSMEs have performed well to date and, if the quality of effort at Rocketdyne stays up to its past standards, should continue to send humans into space reliably and safely.

Space Shuttle Main Engine (SSME) Integrated Subsystem Test Bed (ISTB) rollout, 1975; Left to right: Bob Thompson, NASA SSME Project Manager at MSFC; Norm Revel, Rocketdyne SSME Program Manager; and Dom Sanchini, ISTB Associate Program Manager.

ISTB hot firing on Test Stand A-1 at National Space Technology Laboratories (NSTL), Mississippi.

SSME being installed in SSFL Test Stand A-3.

SSME Hot fire.

Engineering and Test executive staff, 1979; Back row, Left to Right: Barry Waldman, Paul Fuller, Bob Paster, Dane Huang, John Hon, Dick Agulia, Ed Larson, Don Fulton, Frank Lary, and John Chase, front row, seated left to right: Chuck Fargo, Matt EK, Floyd Bennett, and Dick Johnson; not pictured, Howard Griggs, NSTL Engine Test Director.

SSMEs being installed in a Space Shuttle orbiter at Kennedy Space Center (KSC), Florida.

Space Shuttle STS-1 launch, 1981.

SSME cluster post-orbiter landing at Dryden Flight Research Facility (DFRF), California.

Facilities and Equipment preparations for orbiter processing at Northrup Strip, White Sands, NM.

EPILOGUE

The 1940s had seen the assembly at NAA of a core of very talented and focused young engineers and their rapid advancement in the development of new technology for large liquid-propellant rocket engines. Using the A-4 engine for the German V-2 missile as a starting point, they quickly advanced both the thrust level and specific impulse and greatly simplified the system design. Progress in the 1950s was dramatic, with thrust levels climbing to over 1,000,000 lb. The 1960s and 1970s clearly established Rocketdyne as the American leader in rocket engines, with the development and production of engines for the Atlas, Thor, Jupiter, Saturn I, Saturn IB, and Saturn V launch vehicles. The thrust level of the F-1 engine was up to 1,500,000 lb, and the vacuum specific impulse of the hydrogen/oxygen J-2 engine was all the way up to 425 s. The Rocketdyne shop and the test facilities were humming, and employment topped 20,000.

Success had its price, however. After Apollo astronauts Neil Armstrong and Buzz Aldrin landed on the moon on 20 July 1969, American technology, industrial capability, and financial strength were clearly demonstrated to be beyond that of the rival Soviet Union. New large engine projects came to a halt, with the one outstanding exception of the SSME. Winning that was a major triumph for Rocketdyne. For employees within the division, it was all-important to be on that project—if you were not on the SSME, you were probably on the next layoff list. There was even some grumbling against SSME manager Paul Castenholz. An employee had better be on his good side if he wanted to be on his SSME team. Otherwise the employee likely was gone.

For those on the team there were challenges galore. The SSME design was a major departure from Rocketdyne's long dedication to engine simplicity. The engine was far more complex and challenging than anything ever before tackled by Rocketdyne engineers, but triumph they did. Starting in the 1980s hundreds of astronauts, scientists, medical doctors, and even politicians were successfully propelled into Earth orbit by the SSMEs.

The RS-68 Engine

Employment at Rocketdyne surged to 5000 during SSME development but then sagged. Once again there were no large engine contracts in sight. American launch vehicle manufacturers were even selecting Russian rocket engines for their future upgraded launch vehicles. At Rocketdyne, by then a part of the Boeing Company, management decided to develop a new engine using in-house Boeing dollars. The new RS-68 engine, a hydrogen/oxygen design of 650,000 lb of thrust, was aimed at manufacturing simplicity and low cost to help reduce the cost per pound of payload carried to orbit. Its first application was on the Boeing Delta IV launch vehicle. A major factor in minimizing the design and development cost of the RS-68 was the use of the sophisticated computer programs developed during the SSME program. While great for minimizing costs, these programs resulted in minimizing manhours on the project. How could the core team be held intact? What major project was coming next?

No major military projects were on the horizon. For nonmilitary launches the goal was to substantially reduce the cost per pound of delivering payloads to orbit. Rated strictly as a launch vehicle for heavy payloads rather than by the standards for a human transport, the space shuttle was very reliable, but not exactly cheap, and the diminishing fleet of shuttles had been flying since 1981—age was beginning to be a concern. NASA has started and then scrapped several initiatives to design a next-generation launch vehicle that would significantly reduce costs to orbit. In January 2004 President George W. Bush announced a new initiative for NASA that would develop a new launch vehicle suitable for sending humans back to the moon and then on to Mars. The goals were enthusiastically welcomed by NASA personnel, but the vehicle design was not described, and the source of funding was not at all clear.

	Full Power	Minimum Power
Thrust, vac (KN)	3,341	1,922
(K kg-f/Klb-f)	341/751	197/432
Thrust, s/l (KN)	2,918	1,499
(K kg-f/Klb-f)	299/656	153/337
Chamber pressure (MPa)	9.79	5.62
(psia)	1,420	815
Engine mixture ratio	6.0	
I_{sp}, vacuum (sec)	409	
I_{sp}, sea level (sec)	357	

RS-68 operating characteristics.

RS-68 Hotfire.   RS-68 Installed in Data IV Launch vehicle.

The Promising Linear Aerospike

During the years 2000 and 2001, a promising new design aimed at dramatically reducing launch costs was well on its way to flight. The Lockheed Martin "Skunk Works," made famous by Kelly Johnson and Ben Rich, had proposed a radical new design of a fully reusable launch vehicle with the potential, on paper at least, of reaching orbit with a single stage. Such a vehicle would take off vertically, deploy its payload in orbit, return to land like an airplane, be serviced like an airliner, and be ready for its next launch. A key new feature of this design, to be named Venture Star in its production version, was its adaptation of Rocketdyne's Aerospike technology.

The adaptation was really clever. Just like the E-D nozzle had been turned inside out to form the annular Aerospike configuration, for the Venture Star the combustion chambers would be stretched into a linear configuration, expanding their exhaust gas against a cooled ramp. One row of combustion chambers (also known as "thrust cells") and ramp would constitute the upper aft end of the fuselage, and a second row and ramp would form the lower aft end of the fuselage. By differentially throttling the top and bottom outboard segments, full vehicle attitude control could be achieved, in roll as well as in pitch and yaw. The linear Aerospike would of course have the same great specific impulse of the annular Aerospike all the way from sea level to orbit.

The half-scale suborbital version of the Venture Star was called the X-33, and a contract was signed to build and fly it with funding split between NASA and Lockheed Martin. Rocketdyne designed and developed the linear Aerospike engine and labeled it the XRS-2200. Each engine had top and bottom rows of combustion chambers, so that two engines side by side would provide the four quadrants required for thrust vector control on the X-33 without the need for mechanical gimbaling of the engines. Two paired engines were hot fired with very good results at the Stennis Space Center in Mississippi. Throttling capability of all four quadrants demonstrated that the engines could provide pitch, yaw, and roll control for the X-33 as well as the future Venture Star. A pathfinder configuration of the linear Aerospike was also mounted on an SR-71 Blackbird Mach 3 airplane for in-flight demonstration and verification of wind-tunnel results in a supersonic airflow but was deemed not necessary and was canceled before it was ready for flight.

All was going extremely well for the radical new engine. However, the X-33 was designed for liquid hydrogen and oxygen, and the aerodynamically shaped hydrogen tanks needed to have their weight minimized through the use of composite materials. During a pressure test of this tank at MSFC, it ruptured. A decision was made to build a replacement out of metal, but the estimated time to deliver this new tank was over a year. The always ultra-tight NASA budget could not accommodate this slippage, and Lockheed Martin was unable to carry the funding on its own. The X-33, including the SR-71 test flight, came to a halt, which was too bad. Because of creeping weight growth in several subsystems, there were doubts that the ultimate Venture Star could go all the way to orbit with a single stage, but even with a modest booster, throw-away or reusable, it showed great promise. Whether or not it emerges from a new round of studies, the linear Aerospike design is a major step forward in rocket technology and should see future application. With its evolution dating all the way back to 1959 at Rocketdyne, it is about time this advance in technology be put to use.

The winner of the propulsion contract for the next generation launch vehicle will have a promising future. For the other engine developers, the future looks more bleak. There should be a growing national concern about maintaining a liquid-propellant rocket capability in the United States.

L2 Linear Aerospike hot fire in early 1970s.

Dual XRS-2200 Linear Aerospike engine hot fire at Stennis Space Center, Mississippi, 2001.

Artist's concept of Lockheed Martin's X-33 Reusable Launch Vehicle Technology Demonstrator using two XRS-2200 Linear Aerospike engines.

Artist's Concept of Lockheed Martin's VentureStar Single-Stage-to-Orbit Reusable Launch Vehicle using seven RS-2200 Linear Aerospike engines.

Looking Back

If the future path of rocketry in the United States is unclear, the past is full of triumphs. We can look back with pride at the remarkable achievements of an exceptional team of young engineers and technicians who came to be known as Rocketdyne. In a number of fields of advanced technology, there are legendary feats of remarkable achievement. In the history of liquid rocket engine development in America, there is one era that stands out. The successful development of the SSME capped off a fantastic period of rocket engine development at Rocketdyne that saw liquid rocket thrust levels progress from 1500 lb all the way to the 7,500,000 lb from the five engines of the Saturn V booster, and then from throw-away vehicles to the reusable space shuttle. The veteran Bill Ezell summarized the justifiable pride of the Rocketdyners, "As a part of that engineering community for almost four decades, I still marvel at what has been accomplished. During that time, the American reach for space had moved from aspiration to fact. We've made footprints on the moon; now we look toward a permanent presence in space, followed by a return to the Moon and on to Mars."[1]

What led to this remarkable achievement by the team at Rocketdyne? Let us summarize some of the major contributing factors:

1) Dutch Kindelberger and Lee Atwood at North American Aviation exhibited long-range vision, bold management, and a willingness to invest company earnings in building for the future.

2) NAA had earned an outstanding reputation with the Air Force during World War II, which led to the vital technology-advancing contract for the MX-770 Navaho intercontinental missile.

3) Only the brightest, most energetic young engineers were selectively hired. They were then given ever-growing responsibility up to their ability to handle it.

4) Engineers were given a great deal of freedom to try new things, which created a real "engineer's company."

5) The decision to start with the German A-4 rocket engine was important because it led to a very productive working relationship with Wernher von Braun and his Army/NASA team at Huntsville, Alabama, who selected Rocketdyne engines for the Redstone, Jupiter, Saturn I, Saturn IB, Saturn V, and Space Shuttle.

6) Experiencing the power of large rocket engines was exhilarating, which added to the motivation to push ever higher in thrust levels.

7) There was a massive acceleration of effort to beat the Soviet Union to the first ICBM and then the first manned landing on the moon.

8) There was exceptional technical and financial support from the USAF Propulsion Laboratory people at WAFB and EAFB, who were always ready to pursue technology advances, augmented by the generous allocation of Product Improvement funding from BMD.

9) No witch hunting or placing of blame for failures took place. Rather, the emphasis was on team effort to identify and solve problems. Teams were formed based on who could contribute to a solution, without regard to job titles or rank.

All of these factors together are the ingredients that generated the outstanding success of Rocketdyne in developing powerful and reliable rocket engines that enabled humans to set foot on the moon, established the United States as the leader in space exploration, and made it realistic to start seriously planning the journey of humans to the planet Mars.

APPENDICES

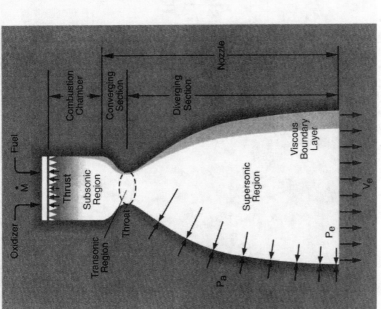

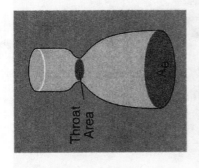

$$\text{Thrust} = (\dot{M} V_e + P_e A_e) - P_a A_e$$

$\dot{M}$ = Engine mass flow rate
V_e = Gas velocity at nozzle exit
P_e = Static pressure at nozzle exit
A_e = Area of nozzle exit
P_a = Ambient pressure

Rocket engine nozzle thrust equation.

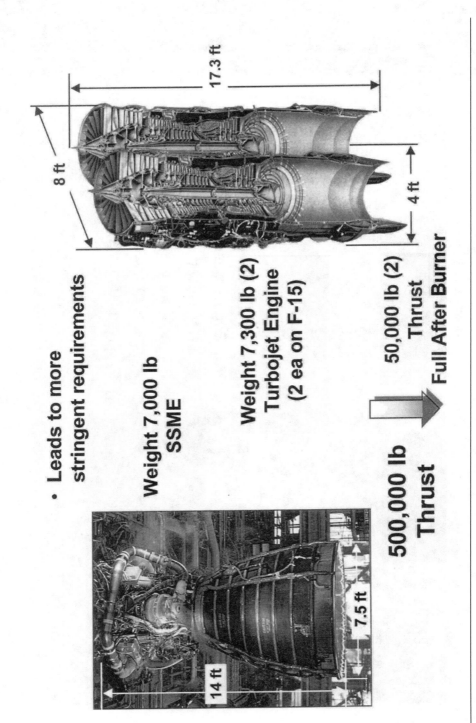

- Leads to more stringent requirements

Weight 7,000 lb
SSME

Weight 7,300 lb (2)
Turbojet Engine
(2 ea on F-15)

500,000 lb Thrust

50,000 lb (2) Thrust
Full After Burner

8 ft
17.3 ft
4 ft
14 ft
7.5 ft

Rocket engines place more demands on the hardware.

Pioneers of Rocketry
#1 of 5
Konstantin Eduardovich Tsiolkovsky (Ziolkovsky) (1857 – 1935)

- Founder of modern science of Astronautics
- Russian of Polish descent
- Teacher of Mathematics and Physics for forty years
- 1895 – 1903 - Published in many scientific journals accounts of space travel-proposed liquid – propellant rocket engines such as hydrocarbons and liquid oxygen
- His designs for rocket-propelled space ships were not blueprints (he did not conduct experiments), but were used to illustrate the principles and difficulties in the conquest of space – included jet vanes in engine exhausts for directional control
- Published many articles on theory of space flight and produced many equations for obtaining required velocity and energy needed for flight to other worlds
- Recognition came late in life when his works were reprinted in the Soviet Union and in 1932, on his 75th birthday, he became a national hero

Konstantin Tsiokovsky

Pioneers of Rocketry
#2 of 5
Dr. Robert H. Goddard (1882 – 1945)

- Father of Modern Rocketry – Pioneer inventor and developer of rockets and missiles
- Started basic research and development which led to the greater expansion of rocket activity during WWII and which eventually lead us to travel beyond Earth's atmosphere
- Career Professor with many technical publications on rocketry had achieved many rocketry "firsts" during his career
- Applied over 200 patents (received 48 before his death and 35 after) on rocket vehicles, engines and components
- Studied and published data on Liquid Hydrogen-liquid oxygen as a propellant combination, but did not test systems with liquid hydrogen as fuel
- Brilliant scientist and inventor – continued to have to struggle for funding – worked experiments in small groups and controlled results closely

Robert H. Goddard

Pioneers of Rocketry
#3 of 5
Professor Hermann J. Oberth (1894 – 1989)

- Known as the founder of European Rocketry
- Classic books on rocketry and astronautics stimulated the active participation of hundreds of other scientists, experimenters, and students in many nations
- Experimented with liquid propellants in Germany – 1929 – 1930 – Member of VfR (German Society for space Travel) – showed the advantages of multistage (step) rockets
- Was right for proposing alcohol-oxygen propellants for lower altitudes, but hydrogen-oxygen for upper stages as later events showed
- Professor at Schassburg and later Mediasch, Translyvania until 1938
- German citizenship took him to Peenemunde starting in 1941 to work on the A-4 and solid rocket missiles until the end of WW II
- Continued his scientific investigations in Germany and Italy post-WW II until coming to the U.S. in 1955
- 1955 - Consultant, Technical Feasibilities Studies Office, Redstone Arsenal untill retirement, when he returned to West Germany

Hermann J. Oberth

Pioneers of Rocketry
#4 of 5
Dr. Theodore von Karman (1881 – 1963)

- Called 'genius's genius' - great theoretician practical scientist, and industrial organizer
- Hungarian Aeronautical Expert – came to the U.S. in 1930
- Responsible for many advances development of high-speed aircraft and rockets
- 1936 - Director of Guggenheim Aeronautical Laboratory at California Institute of Technology (GALCIT Group) program of rocket research and technology (GALCIT later became CalTech's Jet Propulsion Laboratory)
- 1942 – Organized new company known as Aerojet Engineering to work on liquid and soild rocket systems during WW II
- Conducted extensive work with aircraft Jet Assisted Take Off (JATO) units
- 1944 – 1954 – Chairman, Scientific Advisory Board to Chief of Staff, U.S. Air Force
- Authored and co-authored many scientific books, papers, reports, and articles
- Lecturer and member of many technical societies both within the U.S. and internationally

Theodore von Karman

Pioneers of Rocketry
#5 of 5
Dr. Wernher von Braun (1912 – 1977)

- Remarkable scientist, manager and dreamer – it was his eloquent expression of that childhood dream of space travel that made him a national U.S. figure
- 1930 – at age 18, von Braun was student under professor Hermann Oberth – also member of VfR (German society for space Travel)
- 1932 – Hired as German Army's first civilian employee in rocket research – at age 25 was technical director of all such research in Germany
- Presided over development of Germany's V-2 rocket
- 1945 – came to U.S. after WW II as part of "Operation Paperclip"
- His Army Ballistic Missile Agency (ABMA) team launched the first American satellite, Explorer I on 31 January 1958
- Founding Director of NASA's George C. Marshall Space Flight Center – Director July 1960 – February 1970 – the Mercury, Gemini, and Apollo manned spaceflight years
- Spent remaining years at NASA Headquarters performing long range spaceflight planning

Wernher von Braun

Rocketdyne A-6 Redstone

Redstone Ballistic Missile

First Rocketdyne large liquid-propellant engine launch, August 20, 1953.

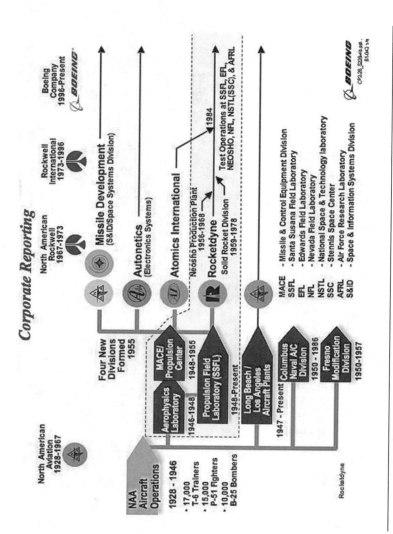

Rocketdyne Heritage.

Appendices 241

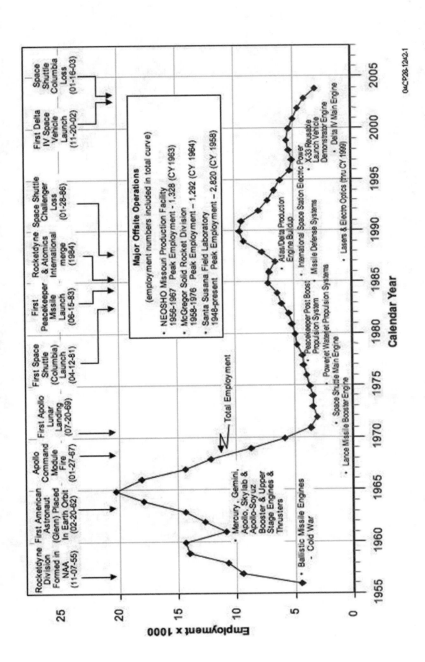

Rocketdyne employment

242 Rocketdyne: Powering Humans into Space

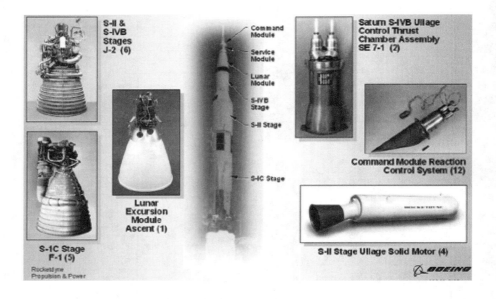

Rocketdyne Engines on Apollo II Flight

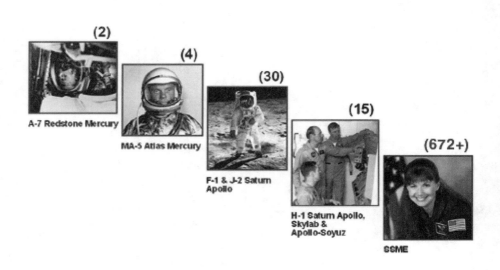

Humans Rocketdyne has powered into space

Appendices 243

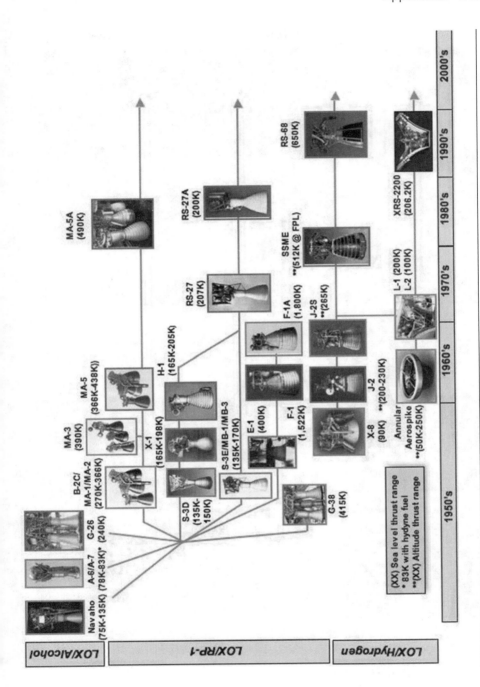

Liquid-propellant propulsion booster evolution

244 Rocketdyne: Powering Humans into Space

Large Launch vehicles and missiles boosted by Rocketdyne engines.

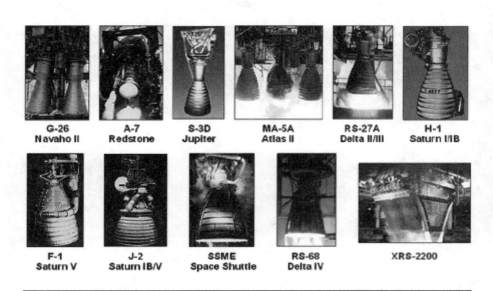

Rocketdyne family of large engines.

Performance Data for Selected Large Rocketdyne Production Engines/Engine Families

Missile/Launch Vehicle Application	Redstone MRBM & SLV	Navaho Cruise Missile	Atlas ICBM & SLV	Jupiter IRBM & SLV	Thor & Delta IRBM & SLV	Saturn I & IB SLV	Saturn V SLV	Saturn IB & V SLV	Space Shuttle SLV	Delta IV SLV
Engine Family Designation (Data from last-in-Family Engine)	A-6 & A-7	G-26	B-2C, MA-1, MA-2, MA-3, MA-5, & MA-5A	S-3D	S-3E, MB-1, MB-3, RS-27, & RS-27A	H-1 A, B, C, & D	F-1	J-2	SSME Phase I & II Block I, IA, IIA, & II	RS-68
Initial Family Engine Design Year	1948	1950	1952	1953	1953	1958	1959	1960	1972	1997
Thrust-Sea Level (Thousand Pounds Force)	78	240	430 (B) 60 (S)	150	200	205	1,522	DNA	374 (100%)	650
-Vacuum	89	278	480 (B) 84 (S)	174	237	237	1,748	230	470 (100%)	745
Specific Impulse (seconds) -Sea Level	218	229	265 (S) 220 (S)	248	255	263	265	DNA	361 (100%)	365
-Vacuum	249	265	295 (B) 309 (S)	288	302	295	305	425	452 (100%)	410
Oxidizer	LOX	LOX	LOX	LOX	LOX	LOX	LOX	LOX	LOX	LOX
Fuel	Alcohol (75%)	Alcohol (92.5%)	RP-1	RP-1	RP-1	RP-1	RP-1	LH_2	LH_2	LH_2
Mixture Ratio (Oxidizer/Fuel)	1.354:1	1.375:1	2.25:1 (B) 2.27:1 (S)	2.4:1	2.24:1	2.23:1	2.27:1	5.5:1	6.03:1	6.0:1
Chamber Pressure (psia)	318	438	719 (B) 756 (S)	527	700	700	982	717	2,747 (100%)	1,460 (100%)
Nozzle Area Ratio (Exit/Throat)	3.61	4.6:1	8:1 (B) 25:1 (S)	8:1	12:1	8:1	16:1	27.5:1	69:1	21.5:1
Nominal Flight Duration (Seconds)	121	65	170 (B) 368 (S)	180	265	150	165	390 (S-II) 580 (IVB)	520	400 Max
Dry Mass (Pounds Mass)	1,478	2,501	3,336 (B) 1,035 (S)	2008	2,528	2,010 (C) 2,041 (D)	18,616	3,454	7,774	14,850
Engine Cycle	GG	GG	GG	GG	GG	GG	GG	GG	SC	GG
Gimbal Angle (Degrees Circular)	None	None	8.5	7.5	8.5	10.5	6	7.5	11.5	10 -MPL 6 -FPL
Diameter/Width (Inches)	68	77	48 (B T/C) 48 (S)	67	67	66	149	81	96	96
Length (Inches)	131	117	101 (B) 97 (S)	142	149	103	230	133	168	205
Operating Temp. Limits (Degrees F)	−25 to +110	−20 to +110	−30 to +130	+40 to +130	0 to +130	0 to +130	−20 to +130	−65 to +140	−20 to +130	−20 to +140
First Flight Date In Family	08–20–1953	11–06–1956	06–11–1957	03–01–1957	01–25–1957	10–27–1961	11–09–1967	02–26–1966	04-12–1981	11–20–2002
Comments		Two Thrust Chamber System	Booster (B) – Two Thrust Chamber System					Restart in Space	Throttleable Power Level 67% to 109%	Throttleable Power Level 57% to 102%

B = Booster, S = Sustainer, GG = Gas Generator Engine Cycle, SC = Staged Combustion Cycle, DNA = Does Not Apply, MPL = Minimum Power Level, FPL = Full Power Level, A&C = Inboard Engines, B&D = Outboard Engines, SLV = Space Launch Vehicle, MRBM = Medium Range Ballistic Missile, IRBM = Intermediate Range Ballistic Missile, ICBM = Intercontinental Ballistic Missile
Table prepared by Vince Wheelock of the Boeing Company, Rocketdyne Propulsion & Power, August 2003

Rocketdyne executive staff photo, 1966; left to right: Doug Hege, George Sutton, Daue Juenke, Wally Fore, Gene Brown, Paul Fritch, Ed Gallant, Sam Hoffman, Unk against wall, Bill Guy, Unk against wall, Joe Mc Namara, Unk, Jack, Armstrong, Paul Vogt, Bob Morin, Bob Thompson, and Bob Lodge.

Appendices **247**

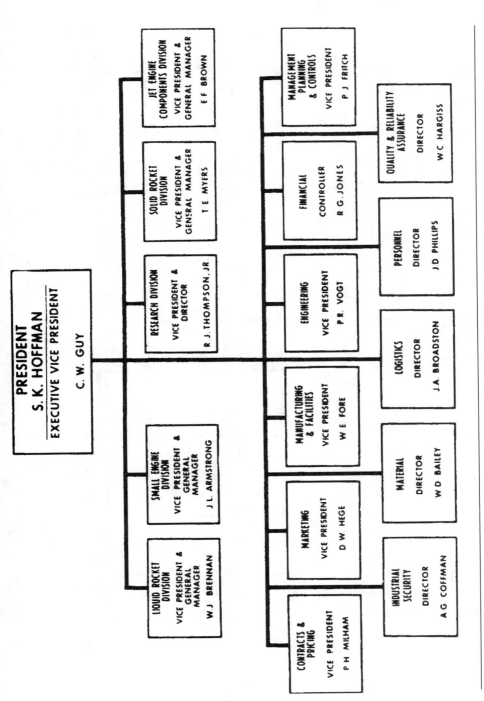

Rocketdyne Division Organization chart during first Apollo launches, 1967.

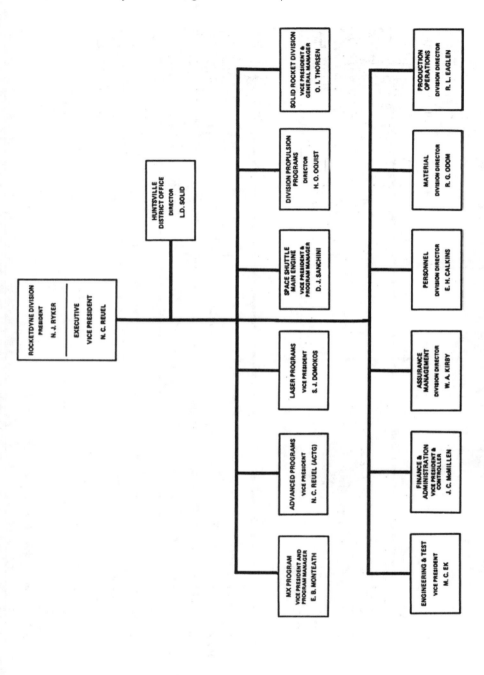

Rocketdyne Division organization chart 10 years after initial Apollo launches, 1977.

Appendices 249

Rocketdyne Canoga Park Aerial view 10 years after landing on Moon, 1979.

Sam Hoffman
General Manager 1955-1960
President 1960-1970

Bill Brennan
President 1970-1976

Norm Ryker
President 1976-1983

Dick Schwartz
President 1983-1989

Robert Paster
President 1989-1994

Paul Smith
President 1994-1997

James Albaugh
President 1997-1998

Russ Turner
Vice President &
General Manager
1998

Byron Wood
Vice President &
General Manager
1998-

Rocketdyne's Presidents and General Managers.

Rocketdyne Santa Susana Field Laboratory Historical Site dedication plaque.

Rocketdyne Santa Susana site designated AIAA Historical site; dedicating personnel left to right: AIAA Vice President Kenneth Sanger, AIAA Chairman Historical Sites Tony Springer, Rocketdyne Deputy Vice President and General Manager John Plowden, and Rockwell International Executive Vice President and Deputy Chairman for Major Programs (retired) Sam Iacobellis.

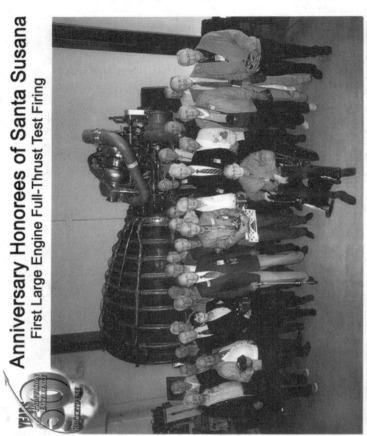

50th Anniversary Honorees, First Full-Thrust Pump-Fed Test Firing in 1950.

NOTES AND REFERENCES

Chapter 1. Rockets, from Theory to the V-2

[1] Ley, W., *Rockets, Missiles, and Space Travel*, Viking Press, New York, 1951, p. 84.
[2] Sears, F. W., *Principles of Physics*, Vol. 1, Addison Wesley Press, Cambridge, MA, 1945, p. 16.
[3] Winter, F. H., *Rockets into Space*, Harvard University Press, Cambridge, MA, 1990, p. 14.
[4] Goddard, R. H., "A Method of Reaching Extreme Altitudes," *Smithsonian Miscellaneous Collections*, Vol. 71, No. 2, *Washington, DC, 1919*.
[5] Oberth, H., *Die Rakete zu den Planetenraumen* (The Rocket into Planetary Space), Verlag von R. Oldenbourg, Munich, 1923.
[6] Winter, F. H., *Rockets into Space*, Harvard University Press, Cambridge, MA, 1990, p. 25.
[7] While both the author and the publisher are committed to the metric system, rocket engines in America have always been identified by their sea-level thrust in pounds, and so in this book we will use metric measures except for thrust, which will be in pounds rather than the metric newtons.
[8] Winter, F. H., *Rockets into Space*, Harvard University Press, Cambridge, MA, 1990, p. 31.
[9] Aerojet History Group, *Aerojet: The Creative Company*, (Stuart F. Cooper, Los Angeles, 1995, pp. 1–18.
[10] Hanrahan, T. H., former Corporal test engineer on the 100K test stand at White Sands Proving Ground, e-mail to R. Kraemer on 18 March 2002.
[11] Heppenheimer, T. A., *Countdown: A History of Space Flight*, Wiley, New York, 1997, p. 51.
[12] Winter, F. H., *Rockets into Space*, Harvard University Press, Cambridge, MA, 1990, p. 46.
[13] Winter, F. H., *Rockets into Space*, Harvard University Press, Cambridge, MA, 1990, p. 50.
[14] Ley, W., *Rockets, Missiles, and Space Travel*, Viking Press, New York, 1951, p. 391.
[15] Ordway, F. I., III, and Sharpe, M. R., *The Rocket Team*, Thomas Y. Crowell, New York 1979.
[16] Neufeld, M. J., *The Rocket and the Reich: Peenemunde and the Coming of the Ballistic Missile Era*, Free Press, New York, 1995.
[17] Neufeld, M. J., *The Rocket and the Reich: Peenemunde and the Coming of the Ballistic Missile Era*, Free Press, New York, 1995, p. 55.

Chapter 2. NAA Chooses the V-2 Path

[1] Jenne, B., *Rockwell; The Heritage of North American Aviation*, Crescent Books, New York, 1989.
[2] *A Short History of North American Aviation, Inc.*, training paper prepared by office of the NAA corporate director for management development, 1958.
[3] *North American History*, pamphlets prepared by The Boeing Company, 1998.
[4] "John Leland Atwood", *Aerospace Management*, June 1963.
[5] "John Leland Atwood", *Forbes*, 15 Sept. 1963.
[6] Heppenheimer, T. A., *Countdown: A History of Space Flight*, Wiley, New York, 1997, p. 57.
[7] Myers, T. F., telephone interview with R. Kraemer, Waco, TX, 3 April 2002.
[8] Tormey, J., telephone interview with R. Kraemer, Pebble Beach, CA, 16 Jan. 2002.
[9] Cecka, W. J., interview with R. Kraemer, Corona Del Mar, CA, 7 Feb. 2002.
[10] Cecka, W. J., Interview with R. Kraemer, Corona Del Mar, CA, 7 Feb. 2002.

254 Rocketdyne: Powering Humans into Space

[11]Winter, F. H., *The East Parking Lot Rocket Experiments of North American Aviation, Inc.*, IAA-99-IAA.2.2.07, 50th International Astronautical Congress, 4–8 Oct. 1999.

[12]"Development of a Strategic Missile and Associated Projects", Aerophysics Laboratory, North American Aviation, Inc., Rept. AL-1347, Oct. 1951.

Chapter 3. Navaho I and Redstone with the 75K Engine

[1]Heppenheimer, T. A., *Countdown: A History of Space Flight*, Wiley, New York, 1997, p. 55.

[2]Neufeld, M. J., Historian at the National Air & Space Museum, e-mail to R. Kraemer, 21 March 2002.

[3]Ezell, W. F., and Mitchell, J. K., "Engine One," *Threshold, Number 7*, Rockwell International, 1991, p. 63.

[4]Heppenheimer, T. A., *Countdown: A History of Space Flight*, Wiley, New York, 1997, p. 55.

[5]Sutton, G. P., interview with R. Kraemer, Los Angeles, CA., 18 Oct. 2001.

[6]Ezell W. F., and Mitchell, J. K., "Engine One," *Threshold, Number 7*, Rockwell International, 1991, p. 59.

[7]Ezell, W. F., and Mitchell, J. K., "Engine One," *Threshold, Number 7*, Rockwell International, 1991, p. 57.

[8]Wheelock, V., "Primal Testing," *Threshold, Number 10,* Rockwell International, 1993, p. 45.

[9]Mitchell, J. K., "Mercury," *Threshold, Number 17*, The Boeing Company, 1998/1999, p. 39.

[10]McCafferty, J. (reference document lost).

[11]Neufeld, M. J., "Orbiter, Overflight, and the First Satellite: New Light on the Vanguard Decision," *Reconsidering Sputnik*, edited by R. D. Launius, J. M. Logsdon, and R. W. Smith, Harwood Academic Publishers, The Netherlands, 2000, Chap. 8.

[12]*NASA Pocket Statistics*, NASA History Office, 1995.

Chapter 4. Navaho II and III and the 120K/135K Engine

[1]"Development of a Strategic Missile and Associated Projects," Aerophysics Laboratory, North American Aviation, Inc., Rept. AL-1347, Oct. 1951.

[2]Dixon, T. F., interview with T. A. Heppenheimer in Brunswick, ME, 4 Feb. 1989, for "The Navaho Program and the Main Line of American Liquid Rocketry," *Air Power History*,1997.

[3]Heppenheimer, T. A., *Countdown: A History of Space Flight*, Wiley, New York, 1997, p. 56.

[4]McCurdy, H. E., *Space and the American Imagination*, Smithsonian Institution Press, Washington, DC, 1997, p. 38.

Chapter 5. The Atlas with 150K and 60K Engines Orbits John Glenn

[1]Yost, M. C., memoirs, *No Riding On or In Rockets or My Thirty Years in a Rocket Factory*, Vol. I, p. 2–7.

[2]Neufeld, J., *The Development of Ballistic Missiles in the United States Air Force 1945-1960*, U.S. Government Printing Office, Washington, DC, 1989, pp. 180–181.

[3]Wheelock, V., interview with R. Kraemer, Westlake, CA, 27 Jan. 2003.

[4]Hege, D. W., personal memoirs, 26 June 2001, p. 13.

[5]Ibid.

[6]Rao, G. V. R., "Exhaust Nozzle Contour for Optimum Thrust," *Jet Propulsion*, Vol. 28, June 1958.

[7]Rao, G. V. R., "Nozzle Contours," *Handbook of Astronautical Engineering*, McGraw-Hill, New York, 1961, Chap. 20.33.

[8]Brennan, W., interview with R. Kraemer, Solana Beach, CA, 16 Nov. 2001.

[9]Kraemer, R. S., *Beyond the Moon; A Golden Age of Planetary Exploration*, Smithsonian Institution Press, Washington, DC, 2000, p. 21.

[10]*Time*, 30 Jan. 1956, p. 54.

[11]Hege, D. W., personal memoirs, 26 June 2001, p. 21.
[12]Heppenheimer, T. A., *Countdown: A History of Space Flight*, Wiley, New York, 1997, p. 131.
[13]Monteath, E., interview with R. Kraemer, Paso Robles, CA, 20 Oct. 2001.
[14]Monteath, E., interview with R. Kraemer, Paso Robles, CA, 20 Oct. 2001.
[15]Mitchell, J. K., "Mercury," p. 43.
[16]Wheelock, V., interview with R. Kraemer, Westlake Village, CA, 27 Jan. 2003.
[17]Monteath, E., e-mail to R. Kraemer, 15 Aug. 2002.
[18]Monteath, E., interview with R. Kraemer, Paso Robles, CA, 20 Oct. 2001.

Chapter 6. H-1 Powered Saturn IB Orbits Apollo 7, Skylab Crews, and Apollo Soyuz

[1]Reuel, N. C., "Liquid Rocket Programs—Review of the Product Line", Rocketdyne, Rept. BC 72-51 26 July 1972.
[2]Godwin, R. (ed.), *X-15, the NASA Mission Reports*, Apogee Books, 2000.
[3]Phillips Petroleum had a solid-propellant plant in MacGregor, TX, producing ammonium nitrate/rubber JATO units for the Air Force, primarily to boost B-47 bombers. The Air Force was unhappy with Phillips because of a poor safety record (ammonium nitrate is potentially explosive) and a corporate reluctance to invest any capital in the plant. With encouragement from the Air Force, NAA purchased a 50% interest in the plant in 1959 and took over management, with Tom Myers as President and John Tormey as his Head of Engineering. Renamed Astrodyne, they developed a much improved propellant binder they called Flexidyne and were a successful business, delivering over 100,000 JATO units and 20,000 Sparrow motors, in addition to Sidewinders and Shrikes, all with a good safety record.
[4]Castenholz, P., interview with R. Kraemer, Ventura, CA, 4 Feb. 2002.
[5]Ezell, W., interview with R. Kraemer, La Quinta, CA, 6 Feb. 2002.
[6]Yost, M. C., personal memoirs, p. 2–3.
[7]Kraemer, R. S., interviewed by M. Neufeld and F. Winter, Washington, DC, 21 Nov. 2000.
[8]Benham, T., telephone interview with R. Kraemer, 27 May 2003.

Chapter 7. Nuclear Rocket Paves Path to Hydrogen/Oxygen J-2 Engine

[1]Gunn, S., *Nuclear Propulsion—A Historical Perspective*, Space Policy 17, Elsevier, 2001, p. 293.
[2]Gunn, S., "The Case for Nuclear Propulsion," *Threshold No. 9*, The Boeing Company, 1992, p. 2.
[3]Mulready, D., *Advanced Engine Development at Pratt & Whitney, 1946 to 1971*, SAE, Warrendale, PA, 2001, p. 58.
[4]Wilhelm, W., interview with R. Kraemer, Woodland Hills, CA, 28 Jan. 2003.
[5]Castenholz, P., interview with R. Kraemer, Ventura, CA, 4 Feb. 2002.
[6]Gunn, S., interview with R. Kraemer, Chatsworth, CA, 8 Feb. 2002.
[7]*The J-2 Rocket Engine*, Rocketdyne Pub. BCI 71-50.
[8]Heppenheimer, T. A., *Countdown: A History of Space Flight*, Wiley, New York, 1997, p. 208.
[9]Castenholz, P., interview with T. A. Heppenheimer, 18 Aug. 1988.
[10]Castenholz, P., interview with R. Kraemer, Ventura, CA, 4 Feb. 2002.
[11]Fenwick, J., "POGO," *Threshold*, Number 8, Rockwell International, 1992, p. 19.

Chapter 8. F-1s and J-2s Power Astronauts to the Moon

[1]Benham, T., telephone interview with R. Kraemer, 27 May 2003.
[2]Baker, D., *Spaceflight and Rocketry: A Chronology*, 1996.
[3]Heppenheimer, T. A., *Countdown: A History of Space Flight*, Wiley, New York, 1997, p. 165.

[4] Biggs, B., interview with R. Kraemer, Canoga Park, CA, 28 Jan. 2003.
[5] Biggs, B., "F-1, The No-Frills Giant," *Threshold, Number 8*, Rockwell International, 1992, p. 24.
[6] Warren, D., and Langer, C. S., *History in the Making—The Mighty F-1 Rocket Engine*, AIAA Paper 89-2387, 10 July 1989, p. 1.
[7] Biggs, B., "F-1, The No-Frills Giant," *Threshold, Number 8*, Rockwell International, 1992, p. 28.
[8] Oefelein, J. C., and Yang, V., "Comprehensive Review of Liquid-Propellant Combustion Instabilities in F-1 Engines," *Journal of Propulsion and Power*, Vol. 9, No. 5, 1993, pp. missing
[9] SSME Phase CD Document RSS-8516, *SSME Design from F-1 and H-1 Experience*, 21 April 1971.
[10] Biggs, B., "F-1, The No-Frills Giant," *Threshold, Number 8*, Rockwell International, 1992, p. 30.
[11] Biggs, B., "F-1, The No-Frills Giant," *Threshold, Number 8*, Rockwell International, 1992, p. 30.
[12] Mitchell, J. K., "Stennis," *Threshold Number 18*, The Boeing Company, 2000, p. 14.
[13] Wheelock, V., interview with R. Kraemer, Westlake Village, CA, 27 Jan. 2003.
[14] Biggs, B., "F-1, The No-Frills Giant," *Threshold, Number 8*, Rockwell International, 1992, p. 31.
[15] Benham, T., telephone interview with R. Kraemer, 27 May 2003.
[16] See Chapter 8 for a discussion of the pogo oscillation experienced on Apollo 13.
[17] Kelly, T. J., *Moon Lander; How We Developed the Apollo Lunar Module*, Grumman Aircraft Company.
[18] Domokos, S., telephone interview with R. Kraemer, 21 March 2002.

Chapter 9. The Fierce Competition for the Space Shuttle Main Engine

[1] Heppenheimer, T. A., "The Space Shuttle Decision", NASA SP-4221, 1999, p. 235.
[2] Iacobellis, S., interview with R. Kraemer, Annapolis, MD, 10 May 2001.
[3] Mulready, D., *Advanced Engine Development at Pratt & Whitney*, p. 58.
[4] Mueller, G., Twenty-fifth Annual Meeting of the International Air Transport Association, 23 Oct. 1969.
[5] Hege, D. W., personal memoirs, 26 June 2001, p. 41.
[6] Biggs, R. E., e-mail to V. Wheelock, "Why We Won the SSME," 5 March 2002.
[7] Benham, T., telephone interview with R. Kraemer, 27 May 2003.
[8] Biggs, B., interview with R. Kraemer, Canoga Park, CA, 28 Jan. 2003.
[9] Mulready, D., *Advanced Engine Development at Pratt & Whitney*, pp. 145–146.
[10] Biggs, B., *Space Shuttle Main Engine; The First Ten Years*, Annual Meeting of the American Astronautical Society, 2 Nov. 1989.
[11] Castenholz, P., interview with R. Kraemer, Ventura, CA, 4 Feb. 2002.
[12] Castenholz, P., interview with T. A. Heppenheimer, 18 Aug. 1988.
[13] Frosch, R., NASA Administrator, Statement to the Senate Subcommittee on Science, Technology and Space, 31 March 1978.
[14] Kraemer, R. S., *Beyond the Moon; A Golden Age of Planetary Exploration*, Smithsonian Institution Press, Washington, DC, 2000, p. 146.
[15] Low, G., personal note No. 119, 27 April 1974.
[16] Biggs, B., "Space Shuttle Main Engine; The First Ten Years,"
[17] Biggs, B., "Space Shuttle Main Engine; The Second Decade", draft manuscript, Jan. 2003. (pages not numbered in the draft copy).
[18] Heppenheimer, T. A., *Development of the Shuttle, 1972-1981, History of the Space Shuttle*, Vol. 2, Smithsonian Institution Press, Washington, DC, 2002, p. 171.
[19] Biggs, B., "Space Shuttle Main Engine; The Second Decade," draft manuscript, Jan. 2003 (pages not numbered in draft copy).
[20] Biggs, B., "Space Shuttle Main Engine; The Second Decade," draft manuscript, Jan. 2003 (pages not numbered in draft copy).

[21] Biggs, B., "Space Shuttle Main Engine; The Second Decade," draft manuscript, Jan. 2003 (pages not numbered in draft copy).
[22] Sutton, G. P., letter to R. S. Kraemer, June 2004.
[23] Sutton, G. P., "History of Liquid-Propellant Rocket Engines in Russia, formerly the Soviet Union", *Journal of Propulsion and Power*, Vol. 19, No. 6, 2003, p. 1024.
[24] Sutton, G. P., exchange of notes with R. S. Kraemer, June and July 2004.
[25] Biggs, B., e-mail to R. S. Kraemer, 20 July 2004.
[26] Biggs, B., e-mail to R. S. Kraemer, 20 July 2004.
[27] Biggs, B., e-mail to R. S. Kraemer, 20 July 2004.
[28] Glenn, J., *Newsweek*, 10 Feb. 2003, p. 39.
[29] Glenn, J., *Newsweek*, 10 Feb. 2003, p. 39.

Epilogue

Ezell, W. F., commentary in *Threshold No. 9*, The Boeing Company, 1992.
For a more complete list of references on liquid-propellant rocket engines, the author highly recommends the books and papers of George P. Sutton, including the following:
Rocket Propulsion Elements, Seventh Edition, coauthored with Oscar Biblarz, Wiley, New York, 2001.
"History of Liquid Propellant Rocket Engines in the United States," *Journal of Propulsion and Power*, Vol. 19, No. 6, 2003, pp.
A History of Liquid Propellant Rocket Engines (to be published by AIAA)

INDEX

A
A-1, A-2, and A-3, *see* rockets, 9
A-3 motor, 10
A-4 chamber, 10
A-4 rocket, 232
  engine, 31-33, 109, 227
  German, 9-11
    derivation of, 58, 163
    engine, 47, 60, 120
  German approach, 36
  see also V-2
A-9 rocket, 29
Abrahamson, Major Gen. James, 218
acetylene, 25-26
Advanced Design Group, 57, 98, 115, 161-162
  advanced development on solid propellant rockets, 122
  drive for simplicity, 120
  led by Floyd Benect, 197
  novel nozzle design, 186
  at Rocketdyne, 83-87, 141, 187, 191
  under Bob Kraemer, 124
  under the leadership of Hege and Sutton, 121
Advanced Design, 57, 127
  managers from, 88
Advanced Research Projects Agency (ARPA), 128, 143-144, 162
AEC *see* Atomic Energy Commission
aerodynamicist, 7
Aerojet Engineering Corporation, 7
Aerojet, 39, 79, 140, 143
  engineering, 22
  motors, 8
  pintle valve assembly, 25
  twin engines designed by, 85
Aerojet's liquid-rocket plant, 85
Aerojet's liquid-rocket program, 73
Aerophysics Laboratory, 22, 29, 35, 57, 87
  Bill Bollay's, 31
    elevating technical capability of, 29
    growing, 29

Aerospace Corporation, 76
aerospace technology, stimulus to, 11
Aerospike engine, 57
  advanced, 98, 121
  design features of annular, 189
  linear, 230
  merits of, 192
Aerospike, 188
  concept, merits of, 191
  gimbaled version, 189
  linear, 229-231
Africano, Alfred, 52
Agulia, Dick, 51, 106, 109, 224
Air Force designation
  "Missile Assembly A" (MA) for Convair, 95, 109
  "Missile Assembly B" (MB) for Douglas, 95
Air Research and Development Command (ARDC), 75-76
aircraft(s)
  AT-6/T-6, Texan advanced trainer aircraft, 18
  B-52 aircraft, 119
  Bell X-1, 7
  DC-1, DC-2, and DC-3 transport, 18
    stress analysis and structural design of, 18
  design, 19
  engines, development of, 162-163
  F-104, 118
  F-86 Saberjet fighter, 115
    FJ-4, navy version, 115
  FJ-1 Fury aircraft, 22
  Lockheed SR-71 Blackbird, 21, 230
  ME-163 interceptor, 9
  P-51 Mustang fighter aircraft, 18-19, 25
  X-15, 21, 118
    dropped from mother B-52 aircraft, 119
    engines for, 120
    rocket powered, 115
  XB-70 bomber, 21

260 Index

Albanese, Phil, 121
alcohol
  as fuel, 36, 73
  pump, 10
  switching to RP-1, 74
  water-diluted, 8
Aldrich, Dave, 74, 161, 169, 171
Aldrin, Buzz, 170, 227
Allison V-1710, 19
alloy of copper and zirconium, 197
Alpha test complex, 94
aluminum, powdered, 122
American Atlas launch vehicles, 221
American Interplanetary Society, 4, 63
American into orbit, first, 101
American rocket engine development companies, 191
American Rocket Society (ARS), 23, 33, 62–63, 89
Ames, Charlie, 73
Anders, Bill, 169
Anders, William, 149
Anderson, Bob, 196
Aniline, 7, 25–26
Apollo 12, 170
Apollo 13, 170
  launch of, 150
Apollo 14–17, 170
Apollo 7, 115
  first manned Apollo spacecraft, 130
Apollo 8, 169
  astronauts on, 149
Apollo lunar missions, 180
Apollo missions 13–17
  impact on moon, 171, 169, 171
Apollo Saturn 501, 169
Apollo Soyuz mission, 147
Apollo Soyuz, 115, 131
Appold, Col. Norman, 143
AR-1, AR-2, AR-3 engines, *see* Engines
Armstrong, Neil, 170
Army Air Corps, 16, 23
Army Air Force (AAF), 25, 29, 33
Army Ballistic Missile Agency (ABMA), 11, 77, 162
  Redstone and Jupiter, 83
Army Corps of Engineers, 89
Army Jupiter IRBM, 125
Army Redstone Arsenal, 43
Arnold Engineering Development Center (AEDC), 147
Arroyo, Seco, 7
Ascent Engine for the Lunar Module (LM), 180
Ashmead, Dick, 60, 84
asphalt binder mix, 7

Association for the Utilization of Industrial Gases (AUIG), 8
Astrodyne, 193
Astrodyne plant, 122
Astronauts
  to Apollo Soyuz, 131
  to skylab 130
  into orbit, H-1 power, 129–137
  on Saturn V, 149
AT-6/T-6, *see* aircrafts, 18
Atlas 60K sustainer engine for RMI engine, 118
Atlas booster engine, LR-89-NA-1, 95
Atlas engines
  designations of, 110
  for Redstone, Navaho II, Navaho III, Thor, and Jupiter, 79
  see also Engines
Atlas F-series test, 106
Atlas ICBM, 77, 95
Atlas III series, with staged combustion Russian RD-180 engines, 221
Atlas missiles, technical development of, 76
Atlas sustainer engine, 83–84
Atlas to orbit, 101–109
Atlas vehicles, 79, 99
  construction at Convair, 94
  deterrent role as an ICBM, 109
  a reliable launch vehicle for a variety of payloads, 109
  first intercontinental guided missile, 82
Atmosphere, composition of, 94
atomic bomb, 38
Atomic Energy Commission (AEC)
  concept of nuclear rocket, 139
  Nevada test site, 140
  Q-clearance from, 74–75
Atomics, International, 80
Atwood, John Leland "Lee", 19–22, 98
  assessment on the results of world war-I exceptional talents of, 16
  effective leader, 31
  elevating technical capability of aerophysics laboratory, 29
  information access from convair, 82
  Kindelberger-Atwood team, 15
  NAA, annual dinner meetings, 87
  personality, 18
  president, NAA, 30
Autonetics division, 80, 95, 193

B

B-52, *see* aircrafts, 119
baffles, injector, 180–181
  addition of, 166

main, and faceplates, 205
  use of, 38, 99
Ballistic missile
  intercontinental, using single stage, 139
  see also Intercontinental Ballistic Missile
Ballistic Missile Division (BMD), 76
balloon tanks, strength of, 95
"Balloon" propellant tanks, 69
balsa powder, 26
Barnsdale, Ross, 200
Bartz, Don, 60
"Basic combat" BC-1 trainer, 17
"Basic trainer" BT-19, 16
  series, success of, 17
Bates, Jim, 123
Bean, Alan, 130, 170
Bebbe, Murray, 35
Becker, Col. Karl Emil, 8, 9
Beech, Ernest, 15
Belew, Lee, 168
Bell Aerospace, 180
Bell Aircraft, 8, 143
bell nozzles, 84, 188
Benham, Theodore "Ted", 162, 169
  on Braun, 129
Bennett, Floyd, 197, 200, 224
Benson, Jim, 37
Berliner-Joyce Aircraft, 15
"Big Engineering," 167
Biggs, Bob, 163
  on Apollo Saturn 501 launching, 169
  a member of F-1 team, 167
  writings
    "Space Shuttle Main Engine; The First Ten Years," 215
    "Space Shuttle Main Engine; The Second Decade," 215
bipropellant gas generator cycle, 60
Blue Streak IRBM, Thor engines for use in, 77
Boattail boots, redesigned, 99
Boattail flow, 98
Boden, O W, 54
Boeing 707, jet-powered, 120
Boeing Company, 228
Bollay, William "Bill," 21–22, 27, 29, 33, 57
bombers
  B-17 and B-24, 19
  B-25 Mitchell, 18
  B-36, 89
Bonestell, Chesley, 62
boost-glide vehicle, 29
Borman, Frank, 149, 169
Bossart, Karel "Charlie," 69–75
  1-1/2 stage ICBM design, 74

balloon tanks, 84
balloon tanks, integrity of, 95
bowl test area, 89, 94
Brand, Vance, 131
Braun, Wernher von, 9, 11, 39, 43, 62, 129
  and his ABMA team, 44
  goals for human space flight, 127, 185
  inspiration of, 62–63
  tram, 77
Bravo test complex, 94
brazing process, 58–60, 95, 165
Brennan, Bill, 84, 167, 194
Bright, Bill, 41
Broadston, Jim, 35
Brown, Don, 200
Brown, Leon, 51
Brown, Powell, 84
BT-19, see "basic trainer"
Bureau of Aeronautics (BuAer), 7, 25
  Power Plant Development Branch, 22
Burlage, Hank, 191
Burry, Roger, 195
Bush, President George W, a new initiative for NASA, 228
Byron, Bob, 74–75

C
Cain, Russ, 51
Canoga Park plant, 44, 165, 193
Canyon test complex, 94
Cape Canaveral, 44, 62, 105, 149
  launching of first Atlas, 96
Carpenter, Scott, 101
Carr, Gerald, 130
Cartoto, Ed, 35
Carvey, Tom, 35
Castenholz, Paul, 23, 36–38, 124
  ad hoc committee on combustion stability 166
  head Experimental Engines group, 122
  opting out of Rocketdyne, 210
  nuclear rocket, 141
  Rocketdyne's program manager on SSME, 196
cavitation, 165, 167
  preventing, 109
Cecka, Bill, 25–26, 29, 35, 85, 87, 88
  in propellant rocket engines, 94
"Centaur", 143
Cernan, Eugene "Gene", 170
Challenger and Columbia failures, 170
Challenger launch, doomed, 216
chamber pressure, 204
Chapman, Colin, 219
Chase, John, 224

Chew, Bernie, 61
chugging, 43
    low-frequency, 37
Clapp, Spencer, 195
Clark, Ross, 171
Coar, Dick, 204
Coca test complex, 94
Collier's magazine, 62, 185
Collins, Mike, 170
Columbia, on mission STS-1, 217
Combs, Paul, 195
combustion chamber, 10, 38, 117
    different shapes, 123
    divided into small chambers, 99
    gas temperature in, 139
    for Venture Star, 229
combustion chamber efficiency,
        demonstration by Pratt & Whitney,
        190
combustion disturbance, 166
combustion efficiency, 26, 188
combustion instability, 37–43, 60, 89, 95, 181
    with hydrogen/ oxygen, 147
combustion pressure, 61
combustion stability, 10, 38, 166
combustor, copper alloy 205
communications satellites, 109
Conrad, Charles "Pete", 130, 170
Consolidated Vultee Aircraft (Convair), 69–75, 82–83, 95
Conyers, Jack, 121, 129
Cook sleds, 91
    rail-mounted, 89
Cook, Ron, 200
coolant velocity, 58
Cooper, Gordon, 101
Corporal missile, 7
Cosmic Ray instrument, James Van Allen's, integration of, 44
Cosmonauts
    Leonov, 131
    Kubasov, 131
Crain, Bob, 211
Crippen, Robert, 217
Cronkite, Walter, 169
Crossfield, Scott, 115, 118
    close calls, 119
Crossland, Doug, 29
cryogenic oxygen and hydrogen lines, 149
Cunningham Walter, 130
Curtiss Flying Service, 15
Curtiss-Wright Corporation, 15
cycles
    FPL certification, 219
    power, 191, 207
    topping, 186

D

Dane, Jack, 35
DC-1, DC-2, and DC-3, *see* aircrafts
Dean, Howard, 76
Degner, Vern, 121
Delta launch vehicle, 77, 136, 228
Delta test complex, 94
Dempsey, Jim, 75
Dennies, Paul, 200
Diem, Hal, 40, 41
diethlyene triamine, 44
Dillaway, Bob, 121
Dillaway, Robert, 140
Dimethylhydrazine, 44
Dingilian, Norm, 168
Disney, Walt, 62
Dixon, Thomas F, 22
Dixon, Tom, 57, 77–80, 122, 123
    pushing for technology advancement, 120
    proposal for million pound thrust engine, 162
    Vice President for development, Rocketdyne, 141
Dodson, Harry, 200
Domokos, Steve, 180
Donovan, Alan, 76
Doolittle, Col. Jimmy, 18
Dornberger, Capt. Walter R, 8–11, 143
Douglas (Thor), 83
Douglas Aircraft, 15, 136, 147
    DC-1 and DC-2, 16
    DC-3, 16, 119
    DC-4 and DC-6, 119
    DC-7s, turbocharged, 119
Douglas Aircraft Company, 16, 18
Douglas space systems center, 135
Douglas, Donald Wills, 16
Downey plant, 29, 43, 96
Dryden flight research facility, 218
Duke, Charles, 170

E

E-1 *see* Engines, Rocketdyne
East Parking Lot test area, 26–29, 34, 71
Eastern airlines, 15
Eastern test range, 62
E-D, *see* Nozzles
E-D nozzle, 229
Edward, Horkey, 19
Edward, S Forman, 7
Edwards Air Force Base (EAFB), 94, 115, 163
    safe landing of Space Shuttle Columbia, 218
Edwards and Holloman Air Force Bases, 89
Edwards Rocket Base (ERB), 163
Ehricke, Krafft, 143
Eisele, Donn, 130

Index **263**

Eisenhower, President Dwight, 73, 98
Ek, Matt, 23, 58, 195
Elks Club, 43
Energiya launch vehicle, RD-0120 engine for, 220
engine(s)
  120K, 120
    all new, 58–62
  135K
    gimbaled, 77
    Thor engine, 85
  150K, 76
    booster engines, 84
    using liquid oxygen for Titan I, 85
  150K Atlas booster engine, 124
  1B H-1, 61
  200K, 145
    hydrogen/oxygen engine (J-2), 144
  60K, 76
  60K Atlas sustainer engine, 84–85
  75K
    Americanized A-4, 33–35
    combustion problems, 37–38
    flight engine for NERVA, 140
    German A-4 derivative of, 58
    heavy-wall construction of, 84
    hydyne-fueled, 45
    powering astronauts into space, 45
    in Redstone missile, 39–43, 89
    testing, 35-37
  Aerospike, linear, 229–231
  AR-1
    development testing of, 118
    and upgrades, 117
  Atlas
    Atlas sustainer engine, new 60K, 83–84
    backup role on Titan, 84–87
    beginning of development, 74–75
    beginning of production, 98
    BMD and Ramo-Wooldridge, organizing, 75–79
    engine designations, 109–113
    development problems, solving, 94–96
    early designs, 70–74
    first launch, a successful failure, 96–100
    Glenn, John rides Atlas to Orbit, 101–109
    living with exploding growth, 87–89
    rapid expansion, 79–83
    Santa Susana Test Facility, expanding, 89–94
    spawning of IRBMs, 76–79
    trimming, 76–79
  Atlas B-2C, 61
  Delta RS-27, 27A, 61
  E-1, 85, 161
    twin 150K Atlas booster engines, 85

F-1
  and J-2, 180
  high-performance hydrogen/oxygen, 204
  instability, solving, 180
  on Saturn V, 130
F-1s and J-2s, powering astronauts to moon
  Apollo ascent engine, rescue by Rocketdyne, 180–183
  Apollo missions to Moon, historic, 169–179
  combustion instability, solving, 166
  designing F-1, 163–166
  enhancing thrust, 161–163
  reliability, 166–169
G-1, Nomad's, 71, 110–111
German A-4, 31–33
GFE V-2, 35
H-1, 125, 127, 131
  massive, 204
  redesignated as RS-27 engines, 136
  with simplifying features of X-1 for Saturn I, 129
hydrogen/ oxygen rocket engine development of, 163
J-2, 130
  need for, 141
J-2, 144, 186, 194, 204
  development of, 146
  on moon, 171
J-2X, hydrogen/oxygen engine, 124
JP-4, 73, 117
Jupiter, 127
MA-1 to 3, 5, and MA-5A, 61
MA-4, proposed single chamber E-1, 109
Monteath alphabetical designation, 110
Navaho, 38–39
Navaho G-26, 38, 61
Navaho, uprating from 120K to 135K, 76
Navaho II and III and the 120K/135K, 57–68
Navaho III booster, propulsion assembly for, 61
Navaho III cluster, 85
nuclear, thermal rocket, 139
Pratt & Whitney (P&W), designations of, 110
Pratt & Whitney RL-10 engine, 127, 135, 190
  Centaur stage, 130, 186
RD-180, Russian, 221
Reaction Motors, 76
Redstone, 39–45

Rocketdyne
  on Apollo II Flight, 242
  for Atlas, Thor, Jupiter, Saturn I, IB, and V launch vehicles, 227
  designations of, 110
  G-38, 39, 61, 62, 76
  G-38 engine package for Navaho III, 76
  large engines, evolution of, 111
  large launch vehicles and missiles boosted by, 244
  large liquid-propellant engine launch, first, 239
  RS-27, 61, 136
  RS-68, 227–229
  RS-68 engine for Boeing Delta IV launch vehicle, 228
  selected large engines/engine families, performance data, 245
Rolls-Royce Merlin, 19
Saturn I, 51, 127
Thor, 77, 144
  MB-1 and −3, 51
twin ramjet engines, 39
with high thrust levels, 227
  hydrogen/oxygen J-2 engine, vacuum specific impulse of, 227
X-1, 124–126, 129
  experimental, 122
  simplified, 163, 204
X-4, 125
XLR-129, 192
XLR and LR series, 35
XLR30, 118
XLR43-NA-1 or 75K, 35
XLR99, 118
XRS-2200, linear Aerospike engine, 230
engine boot for nongimbaled components of, 98
engine efficiency, improving, 34
Engine One, 38
engine reliability J-2, 147
  oxygen/RP-1 H-1 engines, 147
engine simplification
  H-1 for Saturn I and IB, 127–129
    powering astronauts into orbit, 129–137
  role of advanced design, 120–122
  simple design for manned flight, 115–119
  working with Air Force, 119–120
  X-1, experimental engine, pioneering simplicity, 122–127
ethyl alcohol, 26, 34, 58
  substitute for, 44
Evans, Ronald, 170
exhaust velocity, 139
exhaust, hot turbine, 98
expander cycle, 190

expansion efficiency, 84
Expansion-Deflection, concept, 187
Experimental Engines group, 122, 127
  establishment of, 122
  key members of, 123
Explorer 1, 45
Explorer 3, 45
Ezell, Bill, 40, 41, 123

F
F-1, *see* Engines
F-104, *see* Aircrafts
F-86, *see* Aircrafts
Fagan, Bill, 46
Faget, Max, 189, 203
Fairchild camera, 41
Fakler, Lyle, 200
Fargo, Chuck, 224
"Father of the Space Age", 3
Feldman, Alan, 73
Field, Wright, 74
film cooling, 34
Finger, Harry, 140
flame bucket, 38
  pressure pulse from, 109
  water-cooled, 89
flight readiness firing, 217
fluorine, 26
fluorine spillage, 72
fluorine/hydrazine propulsion system, 110
Fluorine/Liquid Oxygen mixture (FLOX), 127
Fokker Aircraft, 15
Fokker, Tony, 15–16
Fons, Phil, 35
Fontaine, Bob, 166
Forbes magazine, 18
Ford Instrument Company, 15
Forman, Edward S, 7
Fort Bliss, 31
Francis, Dick, 123
Frank Malina, 7
Frank Winter, 3, 10
Friedman, Joseph, 41
Frosch Robert, 212
fuel preburner (FPB), 206
fuel pump turbine, 146
"full flow" start, 41
Full Power Level (FPL), 218–219
Fuller, Paul, 39, 144, 188
Fulton, Don, 195
fusion warheads, development of, 73

G
G-1, *see* Engines
G-38, *see* Engines
Garriott, Owen, 130

Index

gas generator cycle, 144, 163, 191–192, 204, 220
gas generator, bipropellant, 60
gaseous fluorine, 71, 73
gasoline, 3
geiger counters, 45
General Electric, 8
General Motors, 15–16
German Messerschmitt ME-109, 19
Gibson, Edward, 130
gimbal
    actuators, 98, 163
    mounting, 76
gimbaling, 136
Glantz, Joe, 72
Glenn L Martin, 16
Glenn, John, 101–109
    Mercury capsule piloted by, 101
    scary moments for, 101, 221
Goddard, Robert, 2–8, 22, 35
Goe, Ron, 195
Gordon, Richard, 170
Gore, Len, 23
Gore, Leonard, 22
Government furnished equipment (GFE), 33
graphite, for reactor core material, 139
graphite-core reactors, 140
graphite jet vanes, 61
Greenfield, Stanley, 23
Griffin, Jim, 74, 75
Griggs, Howard, 224
Grissom, Gus, 45
Grumman, 180
Guggenheim Aeronautical Laboratory, California Institute of Technology (GALCIT), 7
Guggenheim Foundation, 4
guided missiles, 21–22
    see also missiles
Gunn, Stanley, 23, 41, 140–141
Gunther, Fred, 60
Guy, Bill, 88

H

H-1, see Engines
Hahn, Jack, 95
Haise, Fred, 170
Hale, Jim, 215
Hall, Col. Edward, 34, 76, 80, 123
Hamlin, Chan, 39, 77
Hammond, Walt, 51
hardware-rich combustion, 212
Harmon, Tim, 180
Hauenstein, Cliff, 123, 180
H-bomb, 71, 73
Healy, Roy, 23

heat transfer group, 60
heat transfer rates, 37, 58, 204
Heer, Bill, 103
Hege, Douglas W, 22, 39, 47, 74–84, 87, 88, 95
helium, 146
Helmuth Walter Company, 115
Heylandt, Paul, 8
high pressure fuel turbopump (HPFTP), 205–206, 215
    turbine, 219
    turbine blades, 214
high-pressure fuel ducts, 166
Hitler, 11
Hoffman, Mark, 121
Hoffman, Samuel K, 31, 39, 58, 73–74
    determined leader and organizer, 120
    most trusted managers, 87
    as president of Rocketdyne, 80
"Hogans," 29
Holbrook, Bob, 27
Hon, John, 224
Honeywell division, 214
Hoo, Wan, 1
horizontal test stand (HTS), 89
Hot-firing test, 204
Houston, Ed, 180
Huang, Dane, 224
Hughes Aircraft, 76
Hughes, 76, 119
Hummel, Pete, 51
Huntsville, Alabama, 35, 39, 43, 94, 215
    engineers at, 129
    H-1 engines at, 168
    MSFC in, 200
    NASA team at, 232
    team, 162
Huzel, Dieter, 31, 41
hydrazine, 26, 71, 180
    anhydrous, alternate fuel for RP-1, 127
"hydrogen bomb" warheads, 69
hydrogen pump, 144
hydrogen tanks, pressure test of, 230
hydrogen turbopump
    MK15, 146
    for nuclear rocket, 141
Hydyne, 44

I

Iacobellis, Sam, 57, 121, 171, 195
IBM, 148
igniters, liquid and solid, 37
ignition
    sources, 123
    stages of, 118
Inconel X tubes, 165
injector, triplet, 38

injector pattern
  Bell's triplet, 181
  and starting sequences, trial and error testing of, 166
Integrated Subsystem Test Bed (ISTB), 213, 222–223
Intercontinental Ballistic Missile (ICBM), 74, 84
  accurate, 76
  demands of, 83
  design, 73
  development, 77
  program, 75
  see also missiles
Intermediate-range ballistic missile (IRBM)
  Jupiter engine, 77
  technology, 77
  Thor and Jupiter, 76
    design and development, 77
International Space Station, assembly of, 131
ion propulsion, proponents of, 121
Irvine, General, 79–80
Irwin, James, 170

J

J-2, *see* Engines
Jarvis, George, 51
Jeffs, George, 61
jet assisted takeoff (JATO), 7, 22, 25
  liquid, 6
jet engine, hydrogen-fueled, 144
jet exhaust effect, 19
Jet Propulsion Laboratory (JPL), 7–8, 25, 44–45, 60
jet stream, 119
jet vanes, 124
Johnson Space Center (JSC), 203
Johnson, Dick, 224
Johnson, Jerry, 215
Johnson, Kelly, 229
Johnson, Roy, 143
Jolicoeur, Duncan, 29
Jules Verne, 1, 3
Juno-1, 45
Jupiter, 79

K

"Kaiser hat," 219
Kanarek, Irv, 44
Karman, Theodore von, 21
Keeler, Ray, 27
Kel-F liner, to overcome friction in pump, 109
Kellogg, M W, 8

Kennedy Space Center (KSC), 94, 149, 158
Kennedy, President John F
  Presidential speech
    on Apollo program, 127
    to land a man on the moon, 148
kerosene as rocket fuel, 73
Kerwin, Joe, 130
Keys, Clement Melville, 15
Khruschev, Nikita, 74
Kindelberger, James H "Dutch," 22, 95
  CEO, NAA, 30
  elevating technical capability of aerophysics laboratory, 29
  Kindelberger-Atwood team, 15
  NAA, annual dinner meetings, 87
"King Kong," 163
  testing, 166
  thrust chamber, 162
Kisicki, Paul, 38
KIWI, 140
Klute, Dan, 166
Koelle, Heinz Herman, 85, 128, 161
Konstantin Eduardovich Tsiolkovsky, 1
Kraemer, Anne, 89
Kraemer, Bob, 57, 60, 121, 124, 188
  team of, 95
Kraemer, Robert S, 187
Kramer, Al, 84
Kubic, Frank, 200

L

Laika, the dog, 44
Laminar flow wing, 19
Langley Research Center, 162
Larry, Frank, 195
Larson, Ed, 215
Larson, Vern, 121, 195
Lary, Frank, 224
launch vehicle (s)
  delta series, 77
  Energiya, RD-0120 engine for, 220
  Molniya, 220
  multistage, fly-back, 62
  N-1, 130
  proposal of new design of fully reusable, 229
  Saturn launch vehicles, 129
  Saturn I, 128
    H-1-powered, 129
    upgraded to Saturn IB, 130
  Saturn IB
    ever reliable, 131
    H-1 powered, 115
    Rocketdyne-powered, 171

Saturn V, 130
   F-1 for, 62
   F-1- and J-2-powered, first human crew, 169
   Rocketdyne-powered, 171
Saturn I and IB, Saturn V, 148, 186
   stages, 148
   S-IV powered by Pratt & Whitney RL10 engines, 147
   Thor/Delta launch vehicles, 136
   X-33, half-scale suborbital version of Venture Star, 230
   Vanguard, 44
   Venture Star, 229
   Viking, 76
Law of Motion, Sir Isaac Newton's third, 1, 3
Lawrence Livermore Laboratories (LLL), 139-140
Lawrence, Lovell Jr, 7
Lawrence, Robert, 26
Leas, Larry, 51
Leland, John, 16
Levine, Bob, 41
Lewis Research Center, 145
Lewis, Paul, 180
Ley, Willy, 1, 10
Lindbergh, Charles, 4
Linse, Bob, 162
liquid air cycle engine (LACE), 122
liquid fluorine, 26, 127
   and ammonia, 71
   operational missile application of, 73
   tank, 72
liquid hydrogen, 127, 140
   fuel for future missiles and launch vehicles, 141
   with liquid oxygen as propellants, 143
   rocket fuel, 146
   sealing very low density and low viscosity, 144
   substitute for RP-1 kerosene, 147
liquid nitrogen cooling coil, 72
liquid oxygen (LOX), 8, 25-26, 33, 73, 127, 205-206
   consumption of, 94
   pump, 166
   temperatures, 36
liquid ozone, 26
liquid propellants, 7, 23
liquid-propellant propulsion booster evolution, 243
liquid-propellant rocket engine, 161
liquid-propellant rocket, 4, 7-8
   first flight of, 3

liquid rocket
   early development of, 3
   large, fully-fueled, 89
liquid rocketry, competition in the history of, 191
Lockheed Constellation, 119
Lockheed Martin "Skunk Works," 229
Lockheed SR-71 Blackbird, 21, 230
Lodge, Bob, 29, 35
Lokis, 44
Los Alamos Scientific Laboratory (LASL), 139-140
Lousma, Jack, 130
Lovell Lawrence Jr, 7
Lovell, James, 149
Lovell, Jim, 169-170
low-pressure fuel turbopump (LPFTP), 205-206
Low, George, 169, 180
LOX pump impeller, 109, 165
Lucas, Bill, 216
Lunar Excursion Module (LEM), 181
Lunar Module, 170
Lunar Roving Vehicle, 170, 181

M

MA, *see* Air force designation, 95
MA-1, MA-2, MA-3, MA-4, MA-5, *see* engines
main combustion chamber (MCC), 205-206
main engine controller (MEC), 206, 214
Malina Frank, 21, 23
Malkin, Mike, 209
manned missions to Earth orbit and to moon, 162
manned spaceflight program, 171
MAR-M-246, nickel-based super alloy, 215
Marquardt Corporation, 122
Mars
   landing humans on, 63
   manned mission to, 140
   mission, 63
Marshall Space Flight Center (MSFC), 11, 129, 144-149, 177
   engine test at, 167
   test firing of F-1 engine cluster, 168
Martin Aircraft, 84
Martin Metals, 215
Martinez, Ram, 180
Mattingly, Ken, 170
Max Valier, 8
McCafferty, James, 43
McClure, Marshall, 149
McDonnell Aircraft, 217
McNamara, Joe, 29, 79
Mercury capsule, 221

268 Index

Mercury launch, 101
Mercury missions, 101
Mercury, Project, 45
methane, 140
method of characteristics (MOC), 83, 187
Meyers, Tom, 193
Michoud Assembly Facility (MAF), 177
Mir, 131
missile(s)
   boattail, 98
   development, 80
   guided, experts on, 76
   guided, surface-to-surface ballistic, 69
   intercontinental ballistic missile (ICBM), 69
   intermediate-range ballistic missile (IRBM)
     Jupiter engine, 77
     technology, 77
     Thor and Jupiter, design and development, 76–77
   Jupiter missiles, 163
   long range, 38
   MX-770 Navaho, 29–30
   MX-770 Navaho I,
     increased range, 34
     V-2 thrust for, 31
   Navaho
     120K engines, 74
     air-start, 89
     cruise missiles, 69
   Navaho I, 29
     configuration, 38–39
   Navaho II and III
     120K/135K Engine, 57–68
     120K Engine, 58–62
     Wernher von Braun, inspiration of, 62–63
   Navaho II booster, extension of, 74
   Navaho II cruise missile, G-26
   Navaho II or G-26, 39
   Navaho II, first launch of, 62
   Navaho III, 39, 60, 163
     booster for, 61
     cruise missile, G-38
   Navaho III booster
     propulsion assembly for, 61
     three-engine 405K propulsion system for, 85
   pulsejet-powered subsonic cruise, 11
   Redstone missile, 35, 39–43, 61
     into orbit, 43–45
   surface-to-surface ballistic, 39
   Titan, 84
   V-1 cruise, pulsejet-powered, 21
   V-2, 51

   V-2 ballistic missile, rocket-powered, 8, 21
     A-4 engine for, 227
     see also Atlas missiles; Atlas vehicles
Missile and Control Equipment (MACE), 29, 43, 57, 80
Mississippi Test Facility (MTF), 158
   Saturn engine testing at, 213
Mitchell, Edgar, 170
Mitsubishi Heavy Industry, 77
Mittelwerk, Nazi SS underground production facility, 11
MK9, *see* turbopump
Molniya, *see* launch vehicle, 220
Monteath, Ed, 57, 74, 75, 121
   on Mercury launch, 101
Moon
   Apollo missions 13–17 impact on, 171
   first human landings on, 180
   landing of Apollo astronauts Armstrong and Aldrin, 227
Moore, John, 193
Morin, Bob, 77
Morris, Owen, 209
Morris, Perry, 84
Morse, Charlie, 122
motor scooters, for test crews, 94
Mower, Bill, 84
Mueller, George, 169–170, 180
   engines for manned space shuttles, 191
   head, manned space flights, 185
   at NASA headquarters, 148, 189
Mulliken, Stuart "Stu," 162
Mulready, Dick, 144
Munson, Warren, 51
Munyon, Wayne, 200
muzzle velocity, 1
MX-770 Navaho, *see* missiles
Myers, Dale, 26, 61, 148
Myers, Murray, 168
Myers, Tom, 23

N

N-1, *see* launch vehicle
NAA, *see* North American Aviation
NAR, *see* North American Rockwell
NARloy-Z, 197
   Ingots of, 197
NASA, *see* National Aeronautics and Space Administration
NASM, *see* National Air & Space Museum
National Advisory Committee for Aeronautics (NACA), 19, 118
   Lewis Research Center, 162

Index **269**

National Aeronautics and Space
    Administration (NASA), 63
  early days of, 163
  establishment of, 140
  high performance goals for space shuttle
    engines, 185
  manned space flight, 148
  Mississippi Test Facility (MTF), 168
  proposal to Rocketdyne, million-pound
    thrust engine, 162
  space shuttle program requirements
    document, 197
  Stennis Space Center, 94
  viking mars lander mission, 214
National Air & Space Museum (NASM), 130,
    171
National Historical Monument, 171
National Space Technology Laboratories
    (NSTL), 168, 213
Navaho, see Missiles; Engines
"Navahogans," 29
Naval Research Laboratory (NRL), 43
Nave, Les, 200
Navion airplanes, 20, 26
Navy's Bureau of Aeronautics (BuAer),
    187
Neosho plant, 79, 133
Neu, Ed, 58
Neufeld, Michael, 11, 44
Neumann, John von, 76
Nevada Field Laboratory (NFL), 190, 197
Newton's third law of motion, 1
nitric acid, 25–26, 85
nitrogen tetroxide (NTO), 180
nitrogen, gaseous, 94
Nomad's G-1 pressure-fed engine, 71
"Noon Saloon," 120
North American Aviation (NAA), 10, 15, 16,
    39, 69, 148, 227
  aerodynamicists, 19
  aircraft designers, 115
  annual dinner meeting, 87
  conflict within, 82
  Elks, 43
  into guided missile age, 22
  inertial guidance developed by, 95
  MACE Division of, 80
  merger with Rockwell-Standard, 193
  Missile Development Division, 95
  propulsion group, 70
  rocket engine development, success of,
    23
  rocket engineers, 75
  rocket-powered plane, 115
  wind tunnels, 98

North American Rockwell Corporation (NAR),
    193, 204
  space division, 158
North American's Space Information Division,
    158
nozzle(s), 83, 98
  development for nuclear rocket, 141
  expansion-deflection, 187
  exhaust flow, 187
  full-flowing, 147
  hydrogen-cooled, 144
  J-2 engine's, 186
  optimum bell, 84
Nuclear Engine for Rocket Vehicle Application
    (NERVA), 140, 143, 145
nuclear fission, 139
nuclear rocket program, 140
nuclear rocket
  development of, 139–141
  J-2 engine, 144–160
    birth of 144–147
    powering astronauts to moon,
      161–179
    powering S-IVBs into orbit,
      147–148
    solving the surprise shutdown,
      148–150
  selling hydrogen/oxygen, 141–144
"nucleate boiling," 58

O
Oberth, Hermann, "Father of the Space Age,"
    3
Occupational Safety and Health
    Administration (OSHA), 26
"Ocean of Storms," 170
Office of Advanced Research and Technology
    (OART), 186, 191
Office of Naval Research (ONR), 43
Ordway III, Frederick, 11
oxidizer plumbing, 150
oxidizer preburner (OPB), 206
oxidizer pump impeller, 127
oxidizer pump, 146
oxidizer valve, 206
oxygen pump, failure of, 167

P
P-51, see Aircrafts
Palladium, 95
Paris Gun, 8–9
Parsons, John W, 7, 23
Pasadena Rose Bowl, 7
Paster, Bob, 224
payload, 8–9, 43–44

Peenemunde, 11
  von Braun's crew, 10, 143
  designation for V-2 rocket, 35
  documentary film on A-4, 10
  Germans at, 33
  location and greenery, 9
  Riedels of, 32
Pentagon, 141
Petrone, Rocco, 210
Phillips, Gen. Samuel "Sam", 180
Photocon sensors, 41
Pickering, William, 60
Pierce, H Franklin, 7
pillbox, 198
pioneers of rocketry, 237–239
Pitcairn Aviation Company, 15
"Pogo", 150
Pogue, William, 130
Polaris design, solid-propellant, 79
Pollack, Marsh, 200
Poole, Larry, 69
Pratt & Whitney (P&W)
  engines, designations of, 110
  protest, 204
  RL10 Centaur engine, 130
    demonstration of, 146
  SSME contract, 220
  turbojet engine company, 144
preburners, 191
  dual, 196
  progressive tests on, 198
prestart, low-thrust, 37–38
Project Mercury, 45
Project Orbiter proposal, 43
Project Orbiter vehicle, 44
propellant pumps, 204
propellant valves, 165
propellants
  liquid hydrogen, 2
  liquid oxygen, 10
  liquid/solid, 3, 180
  ethyl alcohol, 10
  storable, 85, 180
propulsion, capability in ramjet and turbojet, 22
pump(s)
  centrifugal pumps, single-stage, 34
  development, 60
  explosions, 167
  liquid oxygen (LOX), 146
  oxidizer, 146
  oxygen, failure of, 167
  propellant, 204
  technology, 10
  turbopump design, 23, 34

pump discharge pressures, 146, 204, 220
pump impellers, cavitation of, 38

R

Ramjet (s)
  combustion, 23
  development of, 34
Ramo, Simon, 76
Ramo-Wooldridge Corp. (R-W), 76
Rao optimum contours, 84
Rao, G. V. R, 83
rated power level (RPL), 218
  see also RP-1 fuel
Reaction Motors, Inc. (RMI), 7–8, 58
Reagan, President Ronald, 168
Red Fuming Nitric Acid (RFNA), 7, 25
Redding, Edward, 22
Redstone, see Missiles
Rees, Eberhard, 194
Reidyk, Jake, 200
Remote Test Facility, 27–29
Reuel, Norm, 23, 57, 84, 88
  "Big Engineering," 167
  for engine development problems, 87
Rice, Ray, 29
Rich, Ben, 229
Richtenberg, Bill, 95
Riedel II, Klaus, 32
Riedel, Walter J.H. "Papa," 10, 32
Riedel, Walther, 38, 52, 120
  Riedel, III, 32
RL-10, see Engines, Pratt & Whitney
robot spacecraft, 109
Robot spacecraft Surveyor III, 170
rocket(s)
  A-1, A-2, and A-3, 9
  A-4, 232, see also A-4 rocket
  A-9, 29
  air-to-air, NAKA 1.5-in., 122
  black powder, 1
  boost phase, 101
  definition of, 1
  development in Germany, 8
  engine technology, 76
  from theory to the V-2, 1
    German V-2, technical leap to, 8–13
  fuel RP-1, 73–74
  hydrogen/oxygen combustion, 139
  pioneers, 1–4
  propulsion, 15, 22
  standard features for, 4
  testing, parking lot for, 25
rocket engine(s)
  advancements, 23
  design thrust level, 10

efficiency in space, 1
hardware demands, 236
high pressure, staged combustion, 191
large liquid-propellant, 70
liquid-propellant, 33
   development of, 8
   pump-fed, 37
nozzle thrust equation, 235
Rocket Engine Advancement Program (REAP), 74, 83
   management channel of, 123
   program, 161
rocket exhaust, neutralizing, 72
rocket motors, liquid-propellant, 23
rocket pioneers, 1–4
   pre-world war II, developments, 4–8
"Rocket power", 80
Rocket propulsion group, 57
Rocketdyne Canoga park aerial view, 10 years after landing on moon, 249
Rocketdyne division organization chart
   10 years after initial Apollo launches, 248
   during first Apollo launches, 247
Rocketdyne engineers, main goals of, 163
Rocketdyne engines, *see* Engines
Rocketdyne heritage, corporate reporting, 240
Rocketdyne
   achievements of team,
      major contributing factors for, 232–233
   aerospike design, status and merits of, 192
   American leader in rocket engines, 227
   creation of, 80
   customers for, 83
   demonstration of capabilities, 203
   employment at, 87, 241
      surging, 227
   executive staff photo, 1966, 246
   expanding engine development at, 84
   family of large engines, 244
   formation of, 8
   humans powered into space, 101, 242
   North American's commitment to, 204
   Presidents and General Managers, 250
   Santa Susana field laboratory, historical site dedication plaque, 251
   SSME, major triumph for, 227
   technology advancement efforts, 187
Rocketdyne, Rockwell's Space Division, 150
rocketry in United States, looking back, 232–233
Rolls-Royce, 77
Roosa, Stuart, 170
Rosen, Milton, 43
Rosenbaum, Mort, 73
rough combustion cutoff (RCC), 38, 166

Royce, Major R.S, 46
RP-1 fuel, 73–74, 125, 163
   as actuating fluid, 124
   Rated Power Level (RPL), 218
RS-27, RS-68, *see* Engines
Rudolph, Arthur, 11
Ryan, Cornelius, 62
Ryker, Norm, 210

S

Saberliner, 20
Salyuts, Russian 131
Sanchini Dominic "Dom," 123, 161, 195
Santa Susana Test Facility, expanding, 89–94
satellite(s)
   flying ICBM 98
   putting into orbit, 44
   Soviet, 45
Saturn I, IB, V, *see* Launch vehicles
Schirra, Wally, 101, 130
Schmitt, Harrison "Jack," 170
Schnare, Bill, 76, 123
Schriever, Brigadier General, Bernard, 75–77, 84
Schuman, Bob, 69
Schwartz, Dick, 123, 216
Scientific Advisory Committee, 76
SCORE project, 98
Scott, David, 170
"Sea of Tranquility," 170
Sergeant rockets, 44
Sergei, P Korolev, 4
Sharpe, Mitchell, 11
Sheeline, Randy, 122
Shepard, Alan Jr, 45, 170
Shesta, John 7
short-range artillery rockets, 1
Shuster, Ed, 180
shuttles, diminishing fleet of, 228
Silverstein, Abe, 146, 162
Skylab crews, 115
Skylab, saga of, 130
Slauson Avenue plant, 77, 115
Slayton, Deke, 131
Sloop, John, 145, 163
Smithsonian Institution, 2
smooth full-flow starts, Redstone and Atlas engine, 124
smooth start sequence, 204
"Snow jobs," 167
Sobin, Bob, 188
Society for Spaceship Travel, 4
solid rockets, air-launched, development of, 23
Solid, Lee, 101, 103

solid-propellant artillery rockets, 8
solid-propellant booster motors, seal problems with shuttle, 216
solid-propellant mix, 122
"Sound of freedom," 162
sounding rocket(s), 7, 11
    Aerobee, 44
    solid propellant, 2
Space Nuclear Propulsion Office (SNPO), 140
Space shuttle main engine (SSME), 144
    aerospike concept, advanced, 187–189
    competition for, 149, 185–226
    complexity of, 204–209
    development of, 165
    failures
        major cause of, 221
        spectacular, 211–213
    flight of shuttle, 217–218
    NASA goals, 185–186
    Pratt & Whitney, phase B shootout with, 191–193
    Pratt & Whitney, shock at, 203–204
    preliminary design studies of a new, 191
    problems, beginning, 209–211
    prototype demonstration, 196–203
    reliablilty, 216–217
    solutions, painstaking, 213–215
    staged combustion cycle, proposal of, 194–196
    technically challenging, 220
    thrust chamber, prototype, 197
    top management turnover, forced, 216
    towards high-pressure staged combustion, 189–191
    uprating, 218–222
Space Shuttle
    era of, 218
    forerunner of, 62
    reusable, 63
    STS-5, first "operational", launching of, 218
Space station, 62
Space Technology Laboratories (STL), 76
space travel, new era of, 10
space walk, EVA, 130
spacecrafts
    Explorer 1 and 3, 45
Spaceflight, rocket powered, 2, 3
"Spaghetti motor," 58
Sperry Gyroscope Company, 15
Spike, 188
Sputnik I, 44
Sputnik, 127
Stafford, Tom, 131
"staged combustion cycle," 191

staged combustion engine
    design, 194, 203
    flown using LOX/hydrogen, 220
    preparations for, 196
Stalin, Joseph, 74
Stanley, Al, 27
Stapp, Col. John, 89
static firing tests, 10
"Steer Horn" fuel lines, 215
Stennis Space Center (SSC), 168, 213, 230
Stennis, John, 168
Stewart, Frank, 210
Stiff, R C, 85
stress analysis, 23
Studhalter, Walt, 144
subsynchronous whirl, 214
supersonic aircraft, experimental, 115
"Susie," 87, 89
Sutor, Alois "Al," 121
Sutton, George P, 22–23, 33, 36, 52, 57
    knowledge on Russian rocket engines, 130
Sweigert, John, 170
Sycamore Canyon test stand, 106
synchronous and subsynchronous vibration, 219

T
takeoff thrust, 73, 76
tap-off power cycle, 125, 127
Test and Reliability Equipment (TRE), 89
Texan advanced trainer aircraft
    AT-6/T-6, 18
    T-6/T-28, 29
Theodore von Karman, 7
thermal energy, 19
thermonuclear fusion device, 73
Thiel, Adolph "Dolph," 77
Thiokol, 216
Thomas, Ted, 105
Thompson Products, 76
Thompson, Bob, 209, 222
Thompson, J R, 210, 215
Thompson-Ramo-Wooldridge (TRW), 76
Thomson, Jerry, 166, 180
Thor propulsion system, 125
Thorsen, Ole, 77
"thrust cells," 229
thrust chamber, 4, 36, 45
    120K, 58
    A-4 and 75K, 60
    alternate, gimbaling arrangements, 124
    Atlas vernier, 77
    cooled, 7
    nozzle, uncooled portion of, 217
    redesign of, 33

Index   **273**

tests, 37
tubular, hand brazing of, 95
tubular walled, 84
uncooled ablative, 180
thrust vector control, 76, 230
  schemes, 124
thrust vernier motors, 125
thrust, maximum, 84
Time magazine, 87
Tischler, Del, 191
Titan III design, 129
Titan
  backup role on, 84
  propulsion system, 84
Toftoy, Col. Holger N, 11
Tormey, John, 23, 25–26, 44
torpedoes, hydrogen-peroxide-powered, 23
Trans World Airlines (TWA), 15, 119–120
Transcontinental Air Transport (TAT), 15
Treaty of Versailles, 8
triethylaluminum, 163
triethylboron, 163
Trinan, Terry, 180
Truax Bob, 8, 22
Tsiolkovsky, Konstantin Eduardovich, 3
turbine blades, fatigue failure, 95
turbine exhaust flow, 188
turbine exhaust gases, 163
turbine exhaust, 77, 146
turbine, 146
turbojets, development of, 22
turbomachinery, 50
turbopump, 38, 84, 161, 163, 220
  Mark 3, 61
  MK9, 141, 145
  high-frequency vibration in, 214
  lubrication of, 95
  spinning, 41
  testing, 35
  turbine-driven multistage axial flow pump, MK9, 140
  turning, 124

U
Unden, Bob, 200
United Aircraft Corporation, 204
United States Air Force (USAF), 19–20, 34, 39, 73, 139
  contract for development of Nomad, 71
  for full F-1 engine, award to Rocketdyne, 162
  for liquid hydrogen pumping, 140
  missile technology, development of, 76
  pilots, 89
  strategic goals of, 69

unsymmetrical dimethyl hydrazine (UDMH), 180
USAF Propulsion Laboratory, 119, 161

V
V-2 rocket, 23, 35
  design structure, conservative nature of, 69
  at Nazi SS underground production facility, 11
  propellant combination used by, 26
  in surface-to-surface guided missiles, 29
  technical advances of, 15
  technical leap to, 8–13
  thrust for MX-770 Navaho I, 31
Valentin P. Glushko, 4
Valier, Max, 8–9
valve(s)
  high-tech hot gas, 125
  LOX, 165
  oxidizer, 206
  oxidizer and dual main fuel, 165
  propellant valves, 124, 165
Van Allen Belts, 45
Van Allen, James, 44
Vanderberg Air Force Base, 109
Vanguard, 44
Vehige, Joe, 123
Vengeance Weapon 2 (V-2), 11
Venture Star, 229
Verne, Jules, 1, 3
Vertical Test Stand (VTS), 35, 89
  VTS-1, 37, 39
VfR, Verein fur Raumschefart, 4
Viking computer, 214
Viking, 44, 118
  sounding rocket program, 43
Viking, *see also* Missiles
Virgil "Gus" Grissom, 45
Vogt, Paul, 23, 57, 58, 60, 84, 95, 120
  "Big Engineering," 167
  mutiny against management, 205
von Braun, Wernher, 4, 8, 36, 62–63, 145

W
Wagner, Bill, 188, 195
Waite, Larry, 57
Waldman, Barry, 224
Walter, Thiel, 10
Warhead weight, 76
Warren, Stu, 27
Watts, Keith, 146
Weiss, Hank, 72
Weitz, Paul, 130
Werth, Dick, 51
Western Airlines, 15

Western Development Division (WDD), 76
Wethe, Jay, 193
Wheelock, Vince, 79, 136, 168, 195
White Sands Proving Ground (WSPG), 11, 89
Wieseneck, Hank, 121, 188, 195
Wilhelm, Wilbur "Willie," 77, 144
William, Bollay, 7
Williams, Frank, 51
Willinski, Marty, 121
Willy Ley, 1, 10
Wilson, Bob, 61
wind tunnel
    to study the boattail flow, 98
    tests, 187
Winter, Frank H, 3, 10
Wood, Byron, 192
Wooldridge, Dean, 76
Worden, Alfred, 170
World War I, German Fokker airplanes of, 15
World War II, 18–19, 23
    conclusion of, 15
    developments, pre-, 4–8
    production demands of, 16
    results of, 21

Worth, Weldon, 34, 73
Wright-Patterson Air Force Base (WAFB), 80
    Propulsion Laboratory, 123
Wyld, James H, 33

X

X-1 and X-4, see Engines
X-15, see Aircrafts
X-33, see Launch vehicle
XB-70, see Aircrafts
XRS-2200, see Engines

Y

Yaeger, Chuck, test pilot, 115
Yardley, John, 216
Yost, Max "Mike," 123, 180
Young, Bob, 85
Young, John, 170, 217

Z

Zero-g environment, 62
Zucrow, Maurice, 140
Zucrow's combustion research group, 23
Zuech, metallurgist, 197

Supporting Materials

A complete list of AIAA publications is available at http://www.aiaa.org.